AF829074

Studies in Computational Intelligence

Volume 1198

Series Editor

Janusz Kacprzyk, Polish Academy of Sciences, Warsaw, Poland

Editorial Board

Marco Dorigo, Université Libre de Bruxelles, Bruxelles, Belgium

Andries Engelbrecht, University of Stellenbosch, Stellenbosch, South Africa

Vladik Kreinovich, University of Texas at El Paso, El Paso, TX, USA

Francesco Carlo Morabito, Mediterranea University of Reggio Calabria, Reggio Calabria, Italy

Roman Slowinski, Poznan University of Technology, Poznan, Poland

Yingxu Wang, Schulich School of Engineering, Calgary, AB, Canada

Yaochu Jin, Westlake University, Hangzhou, China

The series "Studies in Computational Intelligence" (SCI) publishes new developments and advances in the various areas of computational intelligence—quickly and with a high quality. The intent is to cover the theory, applications, and design methods of computational intelligence, as embedded in the fields of engineering, computer science, physics and life sciences, as well as the methodologies behind them. The series contains monographs, lecture notes and edited volumes in computational intelligence spanning the areas of neural networks, connectionist systems, genetic algorithms, evolutionary computation, artificial intelligence, cellular automata, self-organizing systems, soft computing, fuzzy systems, and hybrid intelligent systems. Of particular value to both the contributors and the readership are the short publication timeframe and the world-wide distribution, which enable both wide and rapid dissemination of research output.

Indexed by SCOPUS, DBLP, WTI AG (Switzerland), zbMATH, SCImago.

All books published in the series are submitted for consideration in Web of Science.

Bernhard Eidel
Editor

GPT for Python-Coding in Computational Materials Science and Mechanics

From Prompt Engineering to Solutions in Worked-Out Examples

Springer

Editor
Bernhard Eidel
TUBA Freiberg
Freiberg, Germany

ISSN 1860-949X ISSN 1860-9503 (electronic)
Studies in Computational Intelligence
ISBN 978-3-031-85469-9 ISBN 978-3-031-85470-5 (eBook)
https://doi.org/10.1007/978-3-031-85470-5

© The Editor(s) (if applicable) and The Author(s), under exclusive license to Springer Nature Switzerland AG 2025

This work is subject to copyright. All rights are solely and exclusively licensed by the Publisher, whether the whole or part of the material is concerned, specifically the rights of translation, reprinting, reuse of illustrations, recitation, broadcasting, reproduction on microfilms or in any other physical way, and transmission or information storage and retrieval, electronic adaptation, computer software, or by similar or dissimilar methodology now known or hereafter developed.
The use of general descriptive names, registered names, trademarks, service marks, etc. in this publication does not imply, even in the absence of a specific statement, that such names are exempt from the relevant protective laws and regulations and therefore free for general use.
The publisher, the authors and the editors are safe to assume that the advice and information in this book are believed to be true and accurate at the date of publication. Neither the publisher nor the authors or the editors give a warranty, expressed or implied, with respect to the material contained herein or for any errors or omissions that may have been made. The publisher remains neutral with regard to jurisdictional claims in published maps and institutional affiliations.

This Springer imprint is published by the registered company Springer Nature Switzerland AG
The registered company address is: Gewerbestrasse 11, 6330 Cham, Switzerland

If disposing of this product, please recycle the paper.

Preface

In an era where the boundary between artificial intelligence and human capability continues to blur, the advent of advanced models like GPT-4 stands as a testament to the ever-expanding possibilities of technology. This book is dedicated to exploring the fascinating intersection of artificial intelligence, specifically the capabilities of GPT-4, with computational materials science and computational mechanics—fields that are pivotal in shaping the future of engineering. The focus here is not only to unveil the potential of GPT-4 in generating sophisticated Python coding assignments but also to deepen the understanding and application of computational techniques in modern engineering problems.

Unleashing the Potential of GPT-4 in Computational Sciences. The first objective of this book is to present a curated selection of intriguing problems from computational materials science and computational mechanics. These problems are carefully chosen for their relevance to current research and industrial applications and their suitability for showcasing the advanced capabilities of GPT-4 in code generation. From predicting material behavior under various conditions to simulating complex mechanical interactions, the problems serve as a canvas on which GPT-4 paints its solutions, demonstrating not just accuracy but creativity in problem-solving.

Engineering of Prompts: The Art and Science Behind Effective Questioning. At the heart of effectively utilizing GPT-4 lies the 'engineering of prompts'—a structured and methodological approach to formulating queries that elicit the most coherent and comprehensive answers from the model. This book dedicates substantial focus to this art, guiding readers through the nuances of crafting prompts that are clear, precise, and tailored to extract specific outcomes. By mastering this skill, engineers and researchers can leverage AI tools like GPT-4 to their fullest potential, turning vague ideas into precise algorithmic actions and reliable solutions.

Rigorous Code Verification: Ensuring Reliability in AI-Generated Outputs. Trust in AI-generated code demands rigorous verification. This book addresses the critical need for robust testing frameworks to validate the code produced by GPT-4. Through detailed walkthroughs of testing strategies and verification protocols, readers will learn how to ensure that the solutions provided by AI not only work in theory but also perform flawlessly in real-world applications. This part not only enhances the reliability of AI as a tool but also instills confidence in its users.

Why This Book Is Indispensable. For students, graduate engineers, and seasoned researchers, this book serves as a vital resource that illustrates the practical integration of AI in their work. It demystifies the process of interfacing with one of the most sophisticated AI models to date, making cutting-edge technology accessible and applicable. For educators, this book provides a blueprint for teaching computational concepts using AI, enriching the curriculum, and preparing students for a future where AI is a ubiquitous part of problem-solving.

Inspiring the Next Generation. By bridging traditional engineering problems with modern AI technology, this book not only enhances the current educational and professional landscape but also serves to inspire the next generation of engineers and developers. It pushes the boundaries of what can be achieved when human ingenuity combines with artificial intelligence, setting a new standard for innovation and excellence in engineering.

This book is more than just a textbook or a reference manual; it is a gateway to the future of engineering, a tool that empowers its readers to build on the cutting edge of technology. It encourages them to explore, experiment, and excel in their respective fields, equipped with the knowledge and tools to use AI not just as an aid, but as a transformative force in computational science.

This book project originated from a module I introduced into the Master's degree program in Computational Materials Science (CMS) at TU Bergakademie Freiberg (TUBAF) during the summer semester of 2023, followed by a continuation in the summer semester of 2024. The co-authors of this volume are students of the CMS program at TUBAF who participated in the course and developed their individual projects as part of its requirements.

Freiberg, Germany
July 2024

Bernhard Eidel

Contents

Contributors

Tejas Viresh Anvekar M5, IMFD, TUBAF, Freiberg, Germany

Vishal Vijendra Badami M5, IMFD, TUBAF, Freiberg, Germany

Mehdi Bakhshi Zadeh M5, IMFD, TUBAF, Freiberg, Germany

Aditya Deshmukh M5, IMFD, TUBAF, Freiberg, Germany

Bernhard Eidel M5, IMFD, TUBAF, Freiberg, Germany

Fenil Lathiya M5, IMFD, TUBAF, Freiberg, Germany

Rahul Narkhede M5, IMFD, TUBAF, Freiberg, Germany

Aagashram Neelakandan M5, IMFD, TUBAF, Freiberg, Germany

Pradeep Periyasamy M5, IMFD, TUBAF, Freiberg, Germany

Muhammad Saad Qureshi M5, IMFD, TUBAF, Freiberg, Germany

Acronyms

AI	Artificial Intelligence
ANN	Artificial Neural Networks
BC	Boundary Condition
BCC	Body Centered Cubic
BCE	Binary Cross Entropy
BGK	Bhatnagar-Gross-Krook
BVP	Boundary Value Problem
CA	Cellular Automata
CFD	Computational Fluid Dynamics
CFL	Courant-Friedrichs-Lewy condition
CLAHE	Contrast Limited Adaptive Histogram Equalization
CNN	Convolutional Neural Network
CPU	Central Processing Unit
CS	Coordinate System
D2Q9	Two-Dimensional, Nine-Directional
DFT	Density Functional Theory
DL	Deep Learning
erf	Error function
FCC	Face Centered Cubic
FD	Finite Differences
FDM	Finite Difference Method
FDTD	Finite Difference Time Domain
FEM	Finite Element Method
FSL	Fick's Second Law
FTCS	Forward Time Centered Space
FVM	Finite Volume Method
GIF	Graphics Interchange Format
GPT	Generative Pre-trained Transformer
GPU	Graphics Processing Unit
HCP	Hexagonal Close Packed
HEA	High Entropy Alloy

IC	Initial Condition
IDE	Integrated Development Environment
IoU	Intersection over Union
IVP	Initial Value Problem
LBM	Lattice Boltzmann Method
LLM	Large Language Model
MD	Molecular Dynamics
MLP	Multi Layer Perceptron
MSE	Mean Square Error
NLP	Natural Language Processing
NN	Neural Network
OpenCV	Open Source Computer Vision Library
PDE	Partial Differential Equation
PF	Phase Field
PML	Perfectly Matched Layer
RK	Reaction Kinetics
TE	Transverse Electric
TL	Transfer Learning
TM	Transverse Magnetic
ZSL	Zero-Shot Learning

Topics of Computational Materials Science

Generation of Atomic Scale Single Crystals

Pradeep Periyasamy and Bernhard Eidel

Abstract This chapter investigates the coding ability of GPT-4 in the generation of pristine single crystals (Face Centered Cubic (FCC), Body-Centered Cubic (BCC), and Hexagonal Close Packed (HCP)) structures on the atomic scale in arbitrary orientation with respect to a Euclidean reference frame. Beyond, the code enables the insertion of edge and screw dislocations into FCC pristine crystals. The generated crystals in atomic resolution can be used in Molecular Dynamics (MD) or Molecular Statics (MS) simulations. The resultant code is checked for correctness and analyzed to gain deeper insights into GPT-4's behavior. Furthermore, the ability of GPT-4 to incorporate human feedback is also examined by providing corrections to the generated code during the evaluation process.

1 Introduction

A solid is considered a crystal when its atoms are arranged with precise and consistent periodicity in their positions. This organization is made possible through something called a *lattice*, which can be divided into two types: Bravais and non-Bravais.

In a Bravais lattice, all the points in the lattice are the same, meaning the atoms are arranged uniformly throughout the crystal. On the other hand, a non-Bravais lattice has different points in the lattice, indicating that the atoms are not the same everywhere. So, the non-Bravais lattice is referred to *lattice with a basis*.

Figure 1a (a) gives an understanding of the overall structure of a crystal, we can think of it as the combination of two things: the lattice (the organized arrangement of points) and the basis (the specific arrangement of atoms at those points). So, in simpler terms:

$$\textbf{Crystal Structure} = \textbf{Lattice} + \textbf{Basis}$$

P. Periyasamy (✉) · B. Eidel
M[5], IMFD, TUBAF, Freiberg, Germany
e-mail: Pradeep.Periyasamy@student.tu-freiberg.de

B. Eidel
e-mail: bernhard.eidel@imfd.tu-freiberg.de

© The Author(s), under exclusive license to Springer Nature Switzerland AG 2025

B. Eidel (ed.), *GPT for Python-Coding in Computational Materials Science and Mechanics*, Studies in Computational Intelligence 1198,
https://doi.org/10.1007/978-3-031-85470-5_1

This equation helps us grasp the fundamental elements that contribute to the way atoms are organized in a crystal.

Figures 1c–e depict the prevalent arrangement of unit cells observed in a variety of metals. For cubic crystal systems, such as FCC or BCC, the lattice parameter is typically denoted by 'a' and represents the distance between adjacent lattice points along each edge of the cubic unit cell. In other crystal systems, such as hexagonal, there may be multiple lattice parameters that define the dimensions along different crystallographic axes. In Figs. 1c, d, the parameter 'a', and in Fig. 1e, both parameters 'a' and 'c', are referred to as lattice parameters or lattice constants.

Metallic specimens are composed of multiple unit cells repeated in a pattern. However, the alignment between the overall orientation of the specimen and the individual crystals it comprises may not necessarily coincide. An effective approach to describe the crystal's orientation relative to the specimen involves defining two distinct coordinate systems (CS): a crystal CS and a global CS as shown in Fig. 1b. The global (specimen) CS, often used in experiments, describes the orientation of a sample based on its geometry, while the crystal CS in crystallography aligns with the crystal lattice. Rotation matrices $\boldsymbol{R}$ facilitate relating these systems by transforming vectors from one system to another. By applying $\boldsymbol{R}$ to vectors in the global CS, they can be converted to the crystal CS.

Rotations of vectors from global to local are carried out according to (1) and vice versa according to (2)

$$\boldsymbol{x}' = \boldsymbol{R}\boldsymbol{x}\,, \tag{1}$$

$$\boldsymbol{x} = \boldsymbol{R}^T\boldsymbol{x}'\,, \tag{2}$$

where $\boldsymbol{x}$ and $\boldsymbol{x}'$ are vectors in the global and crystal CS, respectively.

However, pristine crystals, characterized by highly ordered repeating atomic patterns, do not undergo plastic deformation and are thus relatively uninteresting. Conversely, real-world materials are predominantly governed by defects, which can be categorized into various dimensions: zero-dimensional (point) defects (such as vacancies, interstitials, and substitutions), one-dimensional (line) defects (such as dislocations), two-dimensional (planar) defects (such as grain boundaries, interfaces, and free surfaces), and three-dimensional (volume) defects (such as pores and voids).

In this section, our focus is directed solely towards the examination of defects stemming from dislocations. Dislocations represent linear or one-dimensional irregularities within a crystal lattice, where the alignment of atoms deviates from the normal arrangement. Dislocations are broadly categorized into two types: edge dislocations and screw dislocations. As illustrated in Fig. 1f, an edge dislocation is characterized by the insertion of an extra portion of a plane of atoms, the edge of which terminates within the crystal. This edge is called a dislocation line abbreviated by $\boldsymbol{\xi}$; for an edge dislocation the Burgers vector $\boldsymbol{b}$ is perpendicular to $\boldsymbol{\xi}$. Within the region around the dislocation line there is some localized lattice distortion. The atoms above the dislocation line in Fig. 1f undergo compression, and those below undergo tension.

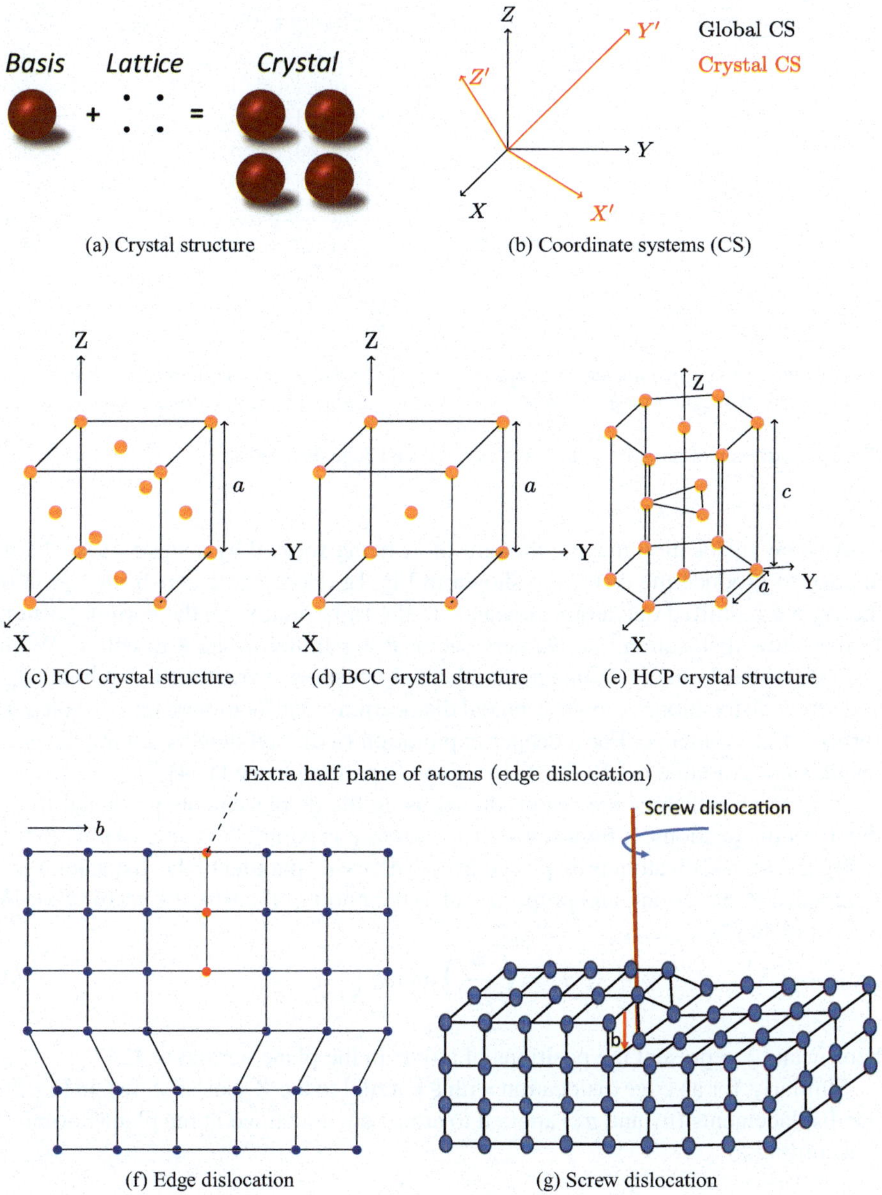

Fig. 1 Definition by sketches of (**a**) crystal structure, (**b**) of coordinate systems, (**c**)–(**e**) of unit cells representing fundamental crystal structures, and representations of (**f**) edge and (**g**) screw dislocations within a crystal lattice

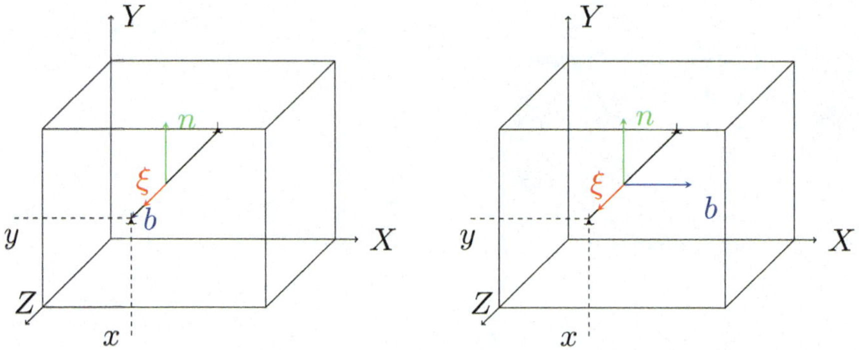

(a) Visualization of parameters involved in screw dislocation for $\boldsymbol{\xi}$ || $\boldsymbol{Z}$-axis

(b) Visualization of parameters involved in edge dislocation for $\boldsymbol{\xi}$ || $\boldsymbol{Z}$-axis

Fig. 2 Illustration of parameters involved in screw and edge dislocation

A screw dislocation may be thought of as being formed by a shear stress that is applied to produce the distortion shown in Fig. 1g. Here, the upper front region of the crystal is shifted one atomic distance to the right relative to the bottom portion. In the screw dislocation, the Burgers vector $\boldsymbol{b}$ is parallel to the $\boldsymbol{\xi}$-direction. While most dislocations in crystalline materials likely possess characteristics of both edge and screw dislocations, known as mixed dislocations, our focus here remains on pure forms for convenience. For a deeper exploration of crystal defects, comprehensive insights are available in [2], for dislocations in particular see [1, 4].

In general, a dislocation is introduced using the displacements predicted in the theory of dislocations. If for a screw dislocations $\boldsymbol{\xi}$ is parallel to the Z-axis as shown in Fig. 2a, then each atom is displaced by a distance u_z parallel to the dislocation line ($\boldsymbol{\xi}$) (hence Z-direction), and proportional to the norm of the Burgers vector $b := |\boldsymbol{b}|$ according to [3]

$$u_z = \left(\frac{b}{2\pi}\right) \arctan\left(\frac{y}{x}\right), \tag{3}$$

here, x and y represent the positions of atoms in the plane normal to $\boldsymbol{\xi}$.

Similarly, for an edge dislocation with $\boldsymbol{\xi}$ parallel to the Z-axis as shown in Fig. 2b, the displacements (u_x and u_y) applied to atoms are contained in the plane normal to ξ according to

$$\begin{aligned} u_x &= \frac{b}{2\pi}\left[\arctan\left(\frac{y}{x}\right) + \frac{xy}{2(1-\nu)(x^2+y^2)}\right], \\ u_y &= -\frac{b}{2\pi}\left[\frac{(1-2\nu)\ln(x^2+y^2)}{4(1-\nu)} + \frac{x^2-y^2}{4(1-\nu)(x^2+y^2)}\right]. \end{aligned} \tag{4}$$

Here, ν is the Poisson's ratio of the material. If the position of $\boldsymbol{\xi}$ is parallel to the X- or Y-axis, the positions of atoms in (3) and (4) have to be adjusted consistently, cf. [3].

In the subsequent sections of this chapter, the prompts generated will assess the Python coding proficiency of GPT-4 in creating single crystals, specifically focusing on the FCC, BCC, and HCP structures in any arbitrary orientations. Additionally, starting from the pristine or defect-free FCC crystal, we will introduce edge and screw dislocations. The visualization of the generated results will be facilitated through OVITO [8], a renowned scientific software extensively employed for visualizing and analyzing molecular and material simulation data in the domain of computational materials science and engineering.

2 Prompt

Users have the flexibility to input a wide range of questions to obtain answers to their queries. However, the key lies in developing the skill to pose questions in a manner that an Artificial Intelligence (AI) engine can comprehend, leading to the production of high-quality and reliable results. An example of such a well-constructed prompt, achieved through prompt engineering, is presented in Fig. 4.

The prompt offers a comprehensive roadmap of the Python program's objectives, detailing specific instructions for generating crystal structures for FCC, BCC, and HCP systems, with an emphasis on accurately determining and saving atom coordinates. Clear function requirements are outlined such as the function name and user inputs. Primitive and basis vectors for each crystal type are explicitly defined to ensure precise calculations for atom positions. The desired output file name is specified, along with the file format identified as XYZ, compatible with visualization tools like OVITO. User interface aspects, including prompts for crystal type, lattice constants, box length with periodicity in each direction, and orientation, are also described. Additionally, suggestions for error-handling measures to validate user inputs, especially for lattice constants and crystal types, are provided. To ensure reproducibility, two key factors are emphasized: clear prompts defining crucial loop adjustments in for loops and a reminder that primitive and basis vectors must be rotated before generating atom positions. The program intentionally excludes from specifying the function for calculating periodicity length. Furthermore, an additional prompt Fig. 3 has been provided to calculate it separately. This decision is made to mitigate the inherent randomness in GPT-4's behavior, ensuring that the code remains reproducible, at least after a few iterations. These topics will be discussed further in detail.

Objective:

- Develop a complete Python script to calculate the periodicity of FCC, BCC, and HCP crystal by the below formula for periodicity.

Function Name:generate_crystal_structure
Inputs Required:

- Crystal Type: Must be one of 'FCC', 'BCC', or 'HCP'.
- Lattice Constant a: Fundamental for all crystal types.
- Additional Lattice Constant c: Required for HCP only, with the condition $c > a$.
- Number of Unit Cells (x, y, z): The number of unit cells in each direction for the simulation box.

Primitive and Basis Vectors:

- FCC: Primitive vectors [[0.5*a, 0.5*a, 0], [0.5*a, 0, 0.5*a], [0, 0.5*a, 0.5*a]], basis vectors [0, 0, 0].
- BCC: Primitive vectors [[-0.5*a, 0.5*a, 0.5*a], [0.5*a, -0.5*a, 0.5*a], [0.5*a, 0.5*a, -0.5*a]], basis vectors [0, 0, 0].
- HCP: Primitive vectors [[0.5*a, -0.5*(3**(1/2))*a, 0], [0.5*a, 0.5*(3**(1/2))*a, 0], [0, 0, c]], basis vectors [[0.5*a, 0.5*(3**(-1/2))*a, 0.25*c], [0.5*a, -0.5*(3**(-1/2))*a, 0.75*c]].

Formulas to Calculate Periodicity Length:

- For FCC & BCC:
  - Periodicity_length = number_of_unit_cells × Lattice_Constant 'a' × norm(primitive_vectors)
- For HCP in x and y direction:
  - Periodicity_length = number_of_unit_cells × Lattice_Constant 'a' × norm(primitive_vectors)
- For z direction in HCP:
  - Periodicity_length = number_of_unit_cells × Lattice_Constant 'c' × norm(primitive_vectors)
- Floor down the output and provide as integers
- Create proper user prompts and error-handling techniques.

Fig. 3 Prompt for calculating periodicity

3 User Input Prompt and Results of Periodicity Calculation

3.1 Prompt for Periodicity Calculation

The user prompt displayed in Fig. 3 showcases the input prompt provided for the periodicity calculation. Additionally, the Python code generated by GPT-4 and the corresponding user inputs and the code output are detailed in code Listing 1.1 and Tables 1 and 2 respectively.

Table 1 Parameters requested from the user and corresponding inputs

Parameters requested	User inputs		
Enter crystal type (FCC, BCC, HCP):	FCC	BCC	HCP
Enter lattice constant 'a' (Å):	3.51	3.3	3.21
Enter lattice constant 'c' (Å):	–	–	5.21
Enter the number of unit cells in each direction for the simulation box (x, y, z):	5, 5, 5	5, 5, 5	5, 5, 5

Table 2 Representation of box length considering periodicity

Description	Code output		
Periodicity lengths (x, y, z):	43, 43, 43	38, 38, 38	51, 51, 135

3.2 Code Listing for Periodicity Calculation

```
import numpy as np

def generate_crystal_structure(crystal_type, a, c, number_of_unit_cells):
    # Validate crystal type
    if crystal_type not in ['FCC', 'BCC', 'HCP']:
        raise ValueError("Crystal type must be 'FCC', 'BCC', or 'HCP'.")

    # Validate lattice constants
    if not a > 0 or (crystal_type == 'HCP' and not c > a):
        raise ValueError("Invalid lattice constants. Ensure 'a' > 0 and for HCP, 'c' > 'a'.")

    # Define primitive vectors
    if crystal_type in ['FCC', 'BCC']:
        primitive_vectors = np.array([[0.5*a, 0.5*a, 0], [0.5*a, 0, 0.5*a], [0, 0.5*a, 0.5*a]])
    elif crystal_type == 'HCP':
        primitive_vectors = np.array([[0.5*a, -0.5*np.sqrt(3)*a, 0], [0.5*a, 0.5*np.sqrt(3)*a, 0], [0, 0, c]])

    # Calculate periodicity
    norm_primitive = np.linalg.norm(primitive_vectors, axis=1)
    if crystal_type in ['FCC', 'BCC']:
        periodicity_length = (np.array(number_of_unit_cells) * a * norm_primitive).astype(int)
    elif crystal_type == 'HCP':
```

```python
        periodicity_length_xy = (np.array(number_of_unit_cells[:2]) * a * norm_primitive[:2]).astype(int)
        periodicity_length_z = (number_of_unit_cells[2] * c * norm_primitive[2]).astype(int)
        periodicity_length = np.concatenate([periodicity_length_xy, [periodicity_length_z]])

    return periodicity_length
def get_user_inputs():
    while True:
        crystal_type = input("Enter crystal type (FCC, BCC, HCP): ").upper()
        if crystal_type in ['FCC', 'BCC', 'HCP']:
            break
        print("Invalid crystal type. Please choose from 'FCC', 'BCC', or 'HCP'.")

    a = float(input("Enter lattice constant 'a': "))

    if crystal_type == 'HCP':
        c = float(input("Enter additional lattice constant 'c' (must be greater than 'a'): "))
    else:
        c = None  # Not required for FCC and BCC

    number_of_unit_cells = input("Enter the number of unit cells in each direction for the simulation box (x,y,z): ")
    number_of_unit_cells = tuple(map(int, number_of_unit_cells.split(',')))

    return crystal_type, a, c, number_of_unit_cells

if __name__ == "__main__":
    try:
        crystal_type, a, c, number_of_unit_cells = get_user_inputs()
        periodicity_length = generate_crystal_structure(crystal_type, a, c, number_of_unit_cells)
        print("Periodicity lengths (x, y, z):", periodicity_length)
    except ValueError as e:
        print("Error:", e)
```

Code Listing 1.1 Code generated by GPT-4 for periodicity calculation

4 User Prompt and Results of Structure Generation with Arbitrary Orientation

4.1 Prompt for Structure Generation

Following the execution of code for periodicity calculation, the user input displayed in Fig. 4 is provided to GPT-4 to generate crystal structures with arbitrary orientations.

Objective:
Develop a complete Python script to calculate and output crystal structures for any arbitrary orientations of Face-Centered Cubic (FCC), Body-Centered Cubic (BCC), and Hexagonal Close Packed (HCP) systems, focusing on precise atom position coordinates generation and saving.

Function name: generate crystal structure
User Inputs Required:

- Crystal Type: Must be one of 'FCC', 'BCC', or 'HCP'.
- Lattice Constant 'a': Fundamental for all crystal types.
- Additional Lattice Constant 'c': Required for HCP only, with the condition $c > a$.
- Periodicity length(x,y,z): The box length with periodicity in each direction for the simulation box.

- Orientation: Defined by a tuple containing three tuples, each for Miller indices in x, y, and z directions, respectively. Give an example of how the user should enter it.

Primitive and Basis Vectors:
These vectors define the initial layout of atoms within a unit cell and it should be rotated based on the given orientation.

- FCC: Primitive vectors [[0.5*a, 0.5*a, 0], [0.5*a, 0, 0.5*a], [0, 0.5*a, 0.5*a]], basis vectors [0, 0, 0].
- BCC: Primitive vectors [[-0.5*a, 0.5*a, 0.5*a], [0.5*a, -0.5*a, 0.5*a], [0.5*a, 0.5*a, -0.5*a]], basis vectors [0, 0, 0].
- HCP: Primitive vectors [[0.5*a, -0.5*(3**(1/2))*a, 0], [0.5*a, 0.5*(3**(1/2))*a, 0], [0, 0, c]], basis vectors [[0.5*a, 0.5*(3**(-1/2))*a, 0.25*c], [0.5*a, -0.5*(3**(-1/2))*a, 0.75*c]].

Rotation Matrix Generation:
Function: rotation_matrix_from_miller_indices(*hkl_x*, *hkl_y*, *hkl_z*)

- Constructs a rotation matrix from Miller indices, essential for orienting the crystal structure.
- Primitive vectors and basis vectors should be rotated before the Atom Position Generation.

Atom Position Generation:

- Important: All positions must be within defined box lengths and checked for validity.

Fig. 4 Prompt for crystal structure generation

Crucial Adjustment - Loop Range for Atom Position Generation:

- The for loop must range between (−Periodicity_length, Periodicity_length+ 1, 1) for FCC and BCC, and adjusted ranges for HCP, ensuring complete coverage of the simulation box.

Position Checks:

- Ensure positions $0 \leq$ atom_position < Periodicity_length + 1.

Output Specifications:

- File Format: XYZ, named `Crystal_structure.xyz`, listing all calculated atom positions.

User Interface:

- Prompt for crystal type, lattice constants, number of unit cells, and orientation, ensuring input validation and error handling, especially for orthogonality of orientation vectors and lattice constants validity.

Error Handling:

- Include checks for input validity, such as crystal types, lattice constants, and orthogonality of orientation vectors.

Fig. 4 (continued)

4.2 Code Listing for Crystal Structure Generation

In code Listing 1.2, the Python code exhibits the conclusive snippet crafted by GPT-4, accompanied by the corresponding user inputs outlined in Table 3. The outcomes of the code execution are visually depicted in Fig. 5.

```
#!/usr/bin/env python
# coding: utf-8

# In[18]:


import numpy as np

# Rotation Matrix Calculation from Miller Indices
def rotation_matrix_from_miller_indices(hkl_x, hkl_y, hkl_z)
    ↪ :
    # Normalize input vectors
    x = np.array(hkl_x) / np.linalg.norm(hkl_x)
    y = np.array(hkl_y) / np.linalg.norm(hkl_y)
    z = np.array(hkl_z) / np.linalg.norm(hkl_z)

    # Ensure orthogonality and compute the rotation matrix
    if np.dot(x, y) == 0 and np.dot(x, z) == 0 and np.dot(y,
    ↪  z) == 0:
```

```
        return np.array([x, y, z]).T
    else:
        raise ValueError("Input Miller indices are not orthogonal.")

# Generate the crystal structure
def generate_crystal_structure(crystal_type, a, c, periodicity_length, orientation):
    rot_matrix = rotation_matrix_from_miller_indices(*orientation)
    positions = []

    if crystal_type == 'FCC':
        primitive_vectors = np.array([[0.5*a, 0.5*a, 0], [0.5*a, 0, 0.5*a], [0, 0.5*a, 0.5*a]])
        basis_vectors = np.array([[0, 0, 0]])
    elif crystal_type == 'BCC':
        primitive_vectors = np.array([[-0.5*a, 0.5*a, 0.5*a], [0.5*a, -0.5*a, 0.5*a], [0.5*a, 0.5*a, -0.5*a]])
        basis_vectors = np.array([[0, 0, 0]])
    elif crystal_type == 'HCP':
        primitive_vectors = np.array([[0.5*a, -0.5*(3**(1/2))*a, 0], [0.5*a, 0.5*(3**(1/2))*a, 0], [0, 0, c]])
        basis_vectors = np.array([[0.5*a, 0.5*(3**(-1/2))*a, 0.25*c], [0.5*a, -0.5*(3**(-1/2))*a, 0.75*c]])
    else:
        raise ValueError("Invalid crystal type.")

    # Apply rotation to primitive and basis vectors
    primitive_vectors = np.dot(primitive_vectors, rot_matrix)
    basis_vectors = np.dot(basis_vectors, rot_matrix)

    # Generate atom positions
    for i in range(-periodicity_length[0], periodicity_length[0] + 1):
        for j in range(-periodicity_length[1], periodicity_length[1] + 1):
            for k in range(-periodicity_length[2], periodicity_length[2] + 1):
                cell_origin = i * primitive_vectors[0] + j * primitive_vectors[1] + k * primitive_vectors[2]
                for basis in basis_vectors:
                    pos = cell_origin + basis
                    if all(0 <= pos[n] <= (periodicity_length[n] + 1) for n in range(3)):
                        positions.append(pos)

    return positions

# Save positions to XYZ file
def save_to_xyz_file(atom_positions):
    with open("Crystal_structure.xyz", "w") as file:
        file.write(f"{len(atom_positions)}\n\n")
        for pos in atom_positions:
```

```python
            file.write(f"Atom {pos[0]} {pos[1]} {pos[2]}\n")

# User Interface
def main():
    crystal_type, a, c, periodicity_length, orientation = get_user_input()
    positions = generate_crystal_structure(crystal_type, a, c, periodicity_length, orientation)
    save_to_xyz_file(positions)
    print("Crystal structure generated and saved to Crystal_structure.xyz.")

def get_user_input():
    # Collect and validate crystal type
    crystal_type = input("Enter crystal type (FCC, BCC, HCP): ").strip().upper()
    while crystal_type not in ['FCC', 'BCC', 'HCP']:
        print("Invalid crystal type. Please enter FCC, BCC, or HCP.")
        crystal_type = input("Enter crystal type (FCC, BCC, HCP): ").strip().upper()

    # Collect and validate lattice constant 'a'
    a = float(input("Enter lattice constant 'a': "))

    # Initialize 'c' for HCP and validate if needed
    c = 0
    if crystal_type == 'HCP':
        c = float(input("Enter additional lattice constant 'c' (must be > a): "))
        while c <= a:
            print("Invalid 'c' value. It must be greater than 'a'.")
            c = float(input("Enter additional lattice constant 'c': "))

    # Collect and validate periodicity length (x, y, z)
    periodicity_length_str = input("Enter periodicity length (x,y,z): ")
    x, y, z = map(int, periodicity_length_str.split(','))

    # Collect and validate orientation
    print("Enter orientation as Miller indices (h,k,l) for x, y, z directions:")
    hkl_x = tuple(map(int, input("Orientation for x direction: ").split(',')))
    hkl_y = tuple(map(int, input("Orientation for y direction: ").split(',')))
    hkl_z = tuple(map(int, input("Orientation for z direction: ").split(',')))

    # Ensure orientation vectors are orthogonal
    # Note: This is a simplified check and may need refinement based on specific requirements
```

```
    if not (np.dot(hkl_x, hkl_y) == 0 and np.dot(hkl_x, hkl_z) == 0 and np.dot(hkl_y, hkl_z) == 0):
        print("Warning: The provided orientation vectors are not orthogonal. Please check your input.")

    return crystal_type, a, c, (x, y, z), (hkl_x, hkl_y, hkl_z)


if __name__ == "__main__":
    main()
```

Code Listing 1.2 Code generated by GPT-4 for single crystal generation

4.3 Results of Structure Generation

The outcomes of generating single crystals of FCC, BCC, and HCP in native orientations are illustrated in Fig. 5a–c as induced by corresponding input parameters listed in Table 3 (Table 4).

Table 3 User inputs and output for crystal structure generation in native orientation

Input requested parameters:	FCC	BCC	HCP
Crystal type (FCC, BCC, HCP):	FCC	BCC	HCP
Lattice constant 'a':	3.51	3.3	3.21
HCP lattice constant 'c' (with c>a):	–	–	5.21
Periodicity length (x, y, z):	43, 43, 43	38, 38, 38	51, 51, 135
Orientation as Miller indices (h, k, l)			
For x direction:	1, 0, 0	1, 0, 0	1, 0, 0
For y direction:	0, 1, 0	0, 1, 0	0, 1, 0
For z direction:	0, 0, 1	0, 0, 1	0, 0, 1
Output: Crystal structure generated and saved to Crystal_structure.xyz			

Table 4 User inputs for structure generation in orientation other than native

Requested input parameters:	FCC	BCC	HCP
Crystal type (FCC, BCC, HCP):	FCC	BCC	HCP
Lattice constant 'a':	3.51	3.3	3.21
HCP lattice constant 'c' (with c>a):	–	–	5.21
Periodicity length (x, y, z):	43, 43, 43	38, 38, 38	51, 51, 135
Orientation as Miller indices (h, k, l)			
For x direction:	1, −1, 2	1, −1, 2	1, −1, 2
For y direction:	−1, 1, 1	−1, 1, 1	−1, 1, 1
For z direction:	1, 1, 0	1, 1, 0	1, 1, 0
Output: Crystal structure generated and saved to Crystal_structure.xyz			

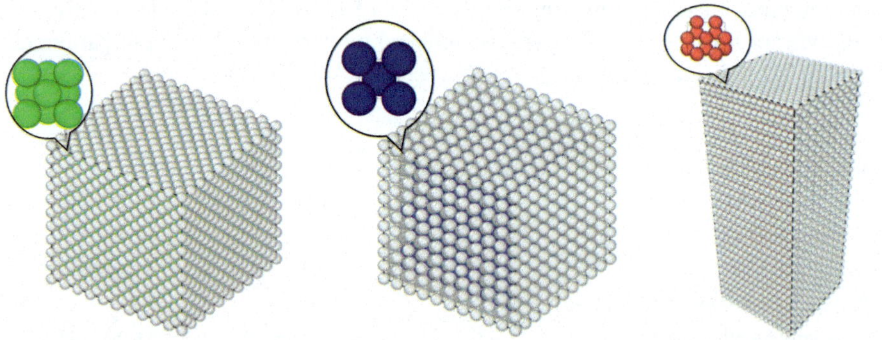

(a) FCC crystal structure (b) BCC crystal structure (c) HCP crystal structure

Fig. 5 Generation of single crystals in native orientation (edges align with < 100 > directions) and visualization using OVITO

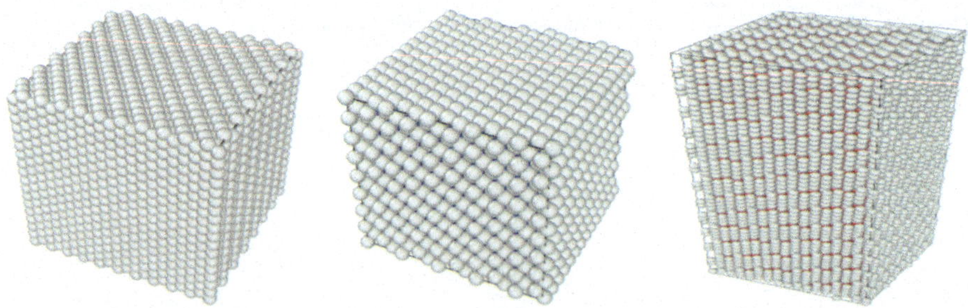

(a) FCC crystal structure (b) BCC crystal structure (c) HCP crystal structure

Fig. 6 Generation of single crystals in orientation other than native and visualization using OVITO

5 User Prompt and Results for Incorporating Dislocations into the Pristine Crystal

5.1 Prompt for Dislocation Insertion

Following the crystal generation process, the user prompt depicted in Fig. 7 was utilized once more to introduce screw and edge dislocations into the pristine FCC samples.

Objective

Please write a Python script that simulates the effect of inserting a dislocation into a crystal structure, modifying the positions of atoms accordingly. The script should include functions to calculate the displacement caused by edge

and screw dislocations in a crystal lattice and apply these displacements to a list of atom positions. It should support dislocations insertion along any of the X, Y, or Z directions.

x1,x2 should be calculated by difference in position between dislocation line(ξ) position and atom position(x,y,z). If ξ=Z then x is the position along X, and y the position along Y; if ξ=Y then z is along Z and x along X; if ξ=X then y is along Y and z along Z.

Function for Edge Displacement:

- Define a function named calculate_edge_displacement that calculates the displacement of atoms due to an edge dislocation.
- It should take the coordinates x1, x2, the Burgers vector magnitude b, and the Poisson's ratio nu as inputs and return the displacement components u1 and u2.

Fig. 7 Prompt for introducing dislocations into the crystal

Function for Screw Displacement:

- Define a function named calculate_screw_displacement that calculates the displacement due to a screw dislocation.
- It takes the coordinates x1, x2, and the Burgers vector magnitude b as inputs, returning the displacement component u3.

'Function to Apply Displacement:

- Create a function named apply_displacement to apply the calculated displacement to each atom's position based on the dislocation type (edge or screw) and direction of the dislocation line (X, Y, or Z) defined by the user.
- The function should check the dislocation type and dislocation direction and then apply corresponding displacements to the correct atom positions.

Function to Insert Dislocation into Crystal:

- Implement a function named insert_dislocation that reads atom positions from an input file in the format of .XYZ, applies the displacement to simulate a dislocation and writes the modified atom positions to an output file in xyz format.
- This function should handle user input to specify the input/output file names, lattice constant, dislocation direction, type of dislocation, Poisson's ratio (if edge dislocation) and dislocation position.

Handling User Inputs:

- At the end of the script, include lines to prompt the user for necessary inputs such as the input and output file names, lattice constant, Poisson's ratio, direction of the dislocation line, and the type of dislocation.
- Based dislocation line direction(X or Y or Z) provided by the user and get the position of dislocation line ($\boldsymbol{\xi}$) for other two coordinates (y,z (if $\boldsymbol{\xi}$ || X) or x,z (if $\boldsymbol{\xi}$ || Y) or x,y (if $\boldsymbol{\xi}$ || Z)) from the user in separate lines. Burgers vector has to be calculated inside the code.

Error Handling:

- Ensure proper error handling in the user inputs as well.

Fig. 7 (continued)

5.2 *Code Listing for Introducing Dislocation*

In code listing 1.3, the Python code exhibits the conclusive snippet crafted by GPT-4, accompanied by the corresponding user inputs outlined in Tables 5 and 6. The outcomes of the code execution are visually depicted in Figs. 8 and 9.

```python
import numpy as np

def calculate_edge_displacement(x1, x2, b, nu):
    theta = np.arctan2(x2, x1)
    r2 = x1**2 + x2**2  # Square of the radial distance r
    u1 = (b / (2 * np.pi)) * (theta + (x1 * x2) / (2 * (1 - nu) * r2))
    u2 = (-b / (2 * np.pi)) * ((1 - 2 * nu) * np.log(r2) / (4 * (1 - nu)) + (x1**2 - x2**2) / (4 * (1 - nu) * r2))
    return u1, u2

def calculate_screw_displacement(x1, x2, b):
    theta = np.arctan2(x2, x1)
    u3 = b * theta / (2 * np.pi)
    return u3

def apply_displacement(atom_positions, dislocation_position, b, nu, direction, dislocation_type):
    modified_atoms = []
    for atom in atom_positions:
        element, x, y, z = atom
        if direction == 'z':
            dx, dy = x - dislocation_position[0], y - dislocation_position[1]
        elif direction == 'x':
            dy, dz = y - dislocation_position[0], z - dislocation_position[1]
        elif direction == 'y':
            dx, dz = x - dislocation_position[0], z - dislocation_position[1]

        if dislocation_type == 'edge':
            if direction == 'z':
                u1, u2 = calculate_edge_displacement(dx, dy, b, nu)
                modified_atoms.append([element, x + u1, y + u2, z])
            elif direction == 'x':
                u1, u2 = calculate_edge_displacement(dy, dz, b, nu)
                modified_atoms.append([element, x, y + u1, z + u2])
            elif direction == 'y':
                u1, u2 = calculate_edge_displacement(dx, dz, b, nu)
                modified_atoms.append([element, x + u1, y, z + u2])
        elif dislocation_type == 'screw':
```

```python
            u3 = calculate_screw_displacement(dx, dy, b)
            if direction == 'z':
                modified_atoms.append([element, x, y, z + u3])
            elif direction == 'x':
                modified_atoms.append([element, x + u3, y, z])
            elif direction == 'y':
                modified_atoms.append([element, x, y + u3, z])
    return modified_atoms

def insert_dislocation(input_file_name, output_file_name, lattice_constant, dislocation_position, direction, dislocation_type, nu):
    with open(input_file_name, 'r') as file:
        lines = file.readlines()

    atoms = [[line.split()[0], float(line.split()[1]), float(line.split()[2]), float(line.split()[3])] for line in lines[2:]]

    b = np.sqrt(2) * lattice_constant / 2  # Burgers vector for FCC

    modified_atoms = apply_displacement(atoms, dislocation_position, b, nu, direction, dislocation_type)

    with open(output_file_name, 'w') as file:
        file.write(f"{len(modified_atoms)}\n")
        file.write(f"Atoms with {dislocation_type} dislocation\n")
        for atom in modified_atoms:
            file.write(" ".join(map(str, atom)) + "\n")

# User input handling
input_file_name = input("Enter the name of the input .xyz file: ")
output_file_name = input("Enter the name of the output .xyz file: ")
lattice_constant = float(input("Enter the lattice constant: "))
nu = float(input("Enter the Poisson's ratio: "))
direction = input("Enter the direction of the dislocation line (x, y, or z): ").lower()
dislocation_type = input("Enter the type of dislocation (screw or edge): ").lower()

# Request additional coordinates based on the direction of the dislocation line
if direction == 'x':
    dislocation_y = float(input("Enter the y-coordinate of the dislocation line: "))
```

```python
    dislocation_z = float(input("Enter the z-coordinate of the dislocation line: "))
    dislocation_position = (dislocation_y, dislocation_z)
elif direction == 'y':
    dislocation_x = float(input("Enter the x-coordinate of the dislocation line: "))
    dislocation_z = float(input("Enter the z-coordinate of the dislocation line: "))
    dislocation_position = (dislocation_x, dislocation_z)
elif direction == 'z':
    dislocation_x = float(input("Enter the x-coordinate of the dislocation line: "))
    dislocation_y = float(input("Enter the y-coordinate of the dislocation line: "))
    dislocation_position = (dislocation_x, dislocation_y)
else:


    raise ValueError("Invalid direction. Pleas enter x, y, or z.")


insert_dislocation(input_file_name, output_file_name, lattice_constant, dislocation_position, direction, dislocation_type, nu)
```

Code Listing 1.3 Code generated by GPT-4 for introducing dislocation into FCC crystal

Table 5 User inputs for insertion of screw dislocation

Requested input parameters:	File names		
Enter the name of the input .xyz file:	Crystal_structure.xyz		
Enter the name of the output .xyz file:	Screw_dislocation.xyz		
Requested input parameters:	Dislocation line		
	X	Y	Z
Enter the lattice constant:	3.51	3.51	3.51
Enter the Poisson's ratio (ν):	0	0	0
Enter the direction of the dislocation line (x, y, or z):	X	Y	Z
Enter the type of dislocation (screw or edge):	Screw	Screw	Screw
Enter the coordinate of the dislocation line:			
For x direction:	–	10	10
For y direction:	10	–	10
For z direction:	10	10	–

Table 6 User inputs for insertion of edge dislocation

Requested input parameters:	File names		
Enter the name of the input .xyz file:	Crystal_structure.xyz		
Enter the name of the output .xyz file:	Edge_dislocation.xyz		
Requested input parameters:	Dislocation line		
	X	Y	Z
Enter the lattice constant:	3.51	3.51	3.51
Enter the Poisson's ratio (ν):	0.33	0.33	0.33
Enter the direction of the dislocation line (x, y, or z):	X	Y	Z
Enter the type of dislocation (screw or edge):	Edge	Edge	Edge
Enter the coordinate of the dislocation line			
For x direction:	–	10	15
For y direction:	10	–	15
For z direction:	10	10	–

5.3 *Results of Dislocation Insertion*

The outcome of the code for inserting screw dislocation and edge dislocation into the crystal is illustrated in Fig. 8. Note that for screw dislocations, the line direction ξ aligns with the Burgers vector.

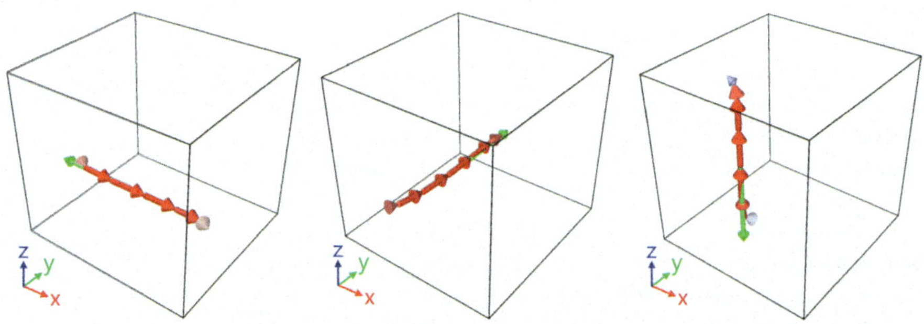

(a) $\boldsymbol{\xi}$-direction || to $\boldsymbol{X}$-axis (b) $\boldsymbol{\xi}$-direction || to $\boldsymbol{Y}$-axis (c) $\boldsymbol{\xi}$-direction || to $\boldsymbol{Z}$-axis

Fig. 8 Screw dislocations in FCC crystals for different orientation. The green arrow indicates the direction of the Burgers vector, the red arrow represents the dislocation line direction. Rendering by using OVITO

The outcome of the code for inserting an edge dislocation or a screw dislocation into the crystal is illustrated in Fig. 9. Note that for edge dislocations the direction ξ of the dislocation line is perpendicular to the Burgers vector.

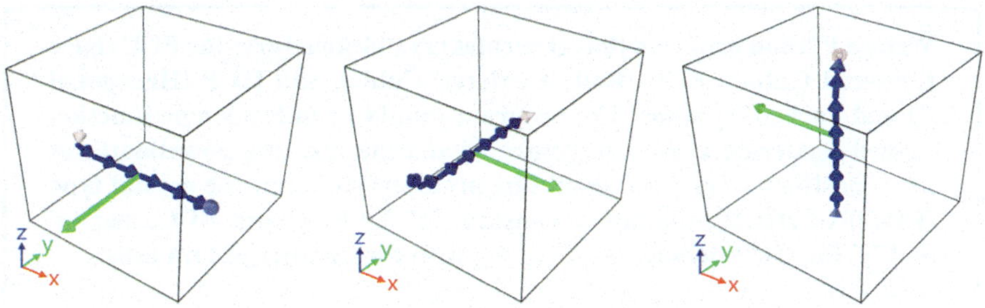

(a) $\boldsymbol{\xi}$ || to $\boldsymbol{X}$-direction (b) $\boldsymbol{\xi}$ || to $\boldsymbol{Y}$-direction (c) $\boldsymbol{\xi}$ || to $\boldsymbol{Z}$-direction

Fig. 9 Edge dislocations in FCC crystals for different orientations. The green arrow indicates the direction of the Burgers vector, while the blue arrows represent the dislocation line direction. Rendering by using OVITO

6 Testing for Verification

6.1 Major Issues

In the initial formulation of the prompt, the problem objective was clearly articulated, as depicted in Fig. 10. GPT-4 demonstrated its proficiency by generating representations of all three crystal structures (FCC, BCC, and HCP) based on this initial prompt. However, a notable observation surfaced during the simulation box depiction, where it became apparent that the atoms did not entirely occupy the designated space, as illustrated in Fig. 11. This discrepancy persisted despite the explicit specification of this condition in the user prompt. The incongruity raises a significant concern regarding the accurate adherence to the specified conditions during the generation process.

To address this issue in the subsequent iteration, a more generic prompt was presented to the language model, as depicted in Fig. 12.

6.1.1 Randomness in the Model

As observed in the subsequent iteration, the introduction of a more generic prompt led to unpredictable behavior in GPT-4. Notably, the model began to deviate from the initially provided user input, altering primitive and basis vectors. Consequently, all three crystal structures generated by the model deviated significantly from the expected configurations, resulting in a failure to detect any valid crystal structures in OVITO as shown in Fig. 13. This outcome underscores the importance of precision in formulating prompts when interacting with Large Language Models (LLMs), as they may exhibit more random behavior in the absence of specific and explicit instructions.

Write a Python program that generates crystal structures for FCC (Face Centered Cubic), BCC (Body Centered Cubic), and HCP (Hexagonal Closed Packed) systems. The program should include a single function named 'generate_crystal_structure' that generates the coordinates of atom positions. The user should be prompted to input the crystal type (FCC/BCC/HCP), the lattice constant 'a' for FCC and BCC, and 'a' and 'c' for HCP (where $c > a$), along with the number of unit cells.

Next, the positions of atoms should be calculated using primitive vectors and basis vectors. Furthermore, the primitive vectors and basis vectors define the arrangement of atoms within the crystal structure and depend on crystal type and lattice constants. The primitive and basis vectors for each crystal are provided below:

For FCC crystal structure:

- Primitive vectors: $[[(1/2) \cdot a, (1/2) \cdot a, 0], [(1/2) \cdot a, 0, (1/2) \cdot a], [0, (1/2) \cdot a, (1/2) \cdot a]]$
- Basis vectors: $[0, 0, 0]$

For BCC crystal structure:

- Primitive vectors: $[[-(1/2) \cdot a, (1/2) \cdot a, (1/2) \cdot a], [(1/2) \cdot a, -(1/2) \cdot a, (1/2) \cdot a], [(1/2) \cdot a, (1/2) \cdot a, -(1/2) \cdot a]]$
- Basis vectors: $[0, 0, 0]$

For HCP crystal structure:

- Primitive vectors: $[[(1/2) \cdot a, -(1/2) \cdot (3^{1/2}) \cdot a, 0], [(1/2) \cdot a, (1/2) \cdot (3^{1/2}) \cdot a, 0], [0, 0, c]]$
- Basis vectors: $[[(1/2) \cdot a, (1/2) \cdot (3^{-1/2}) \cdot a, (1/4) \cdot c], [(1/2) \cdot a, -(1/2) \cdot (3^{-1/2}) \cdot a, (3/4) \cdot c]]$

Explicit Instructions:-

- The simulation box should be completely filled with atoms.
- Also, the code should ensure that the generated atom positions are within the box dimensions, and the final values should be saved in an XYZ file format for visualization in OVITO.

Fig. 10 Initial prompt

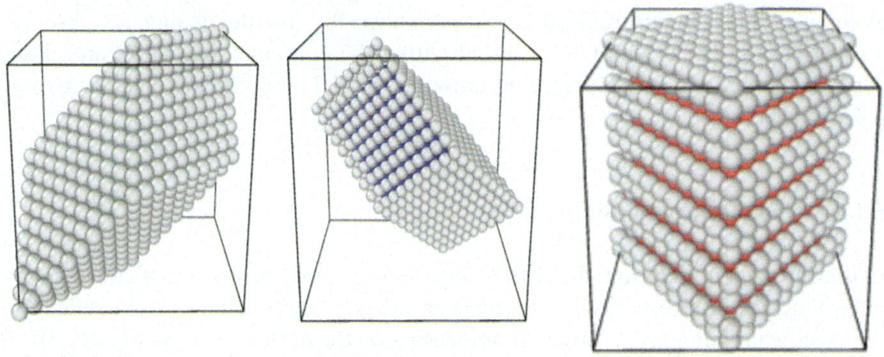

(a) FCC crystal structure (b) BCC crystal structure (c) HCP crystal structure

Fig. 11 Simulation box containing atoms with partial occupancy

The simulation box is not completely filled with atoms

Fig. 12 Iteration 2 to overcome the issue of partial occupancy

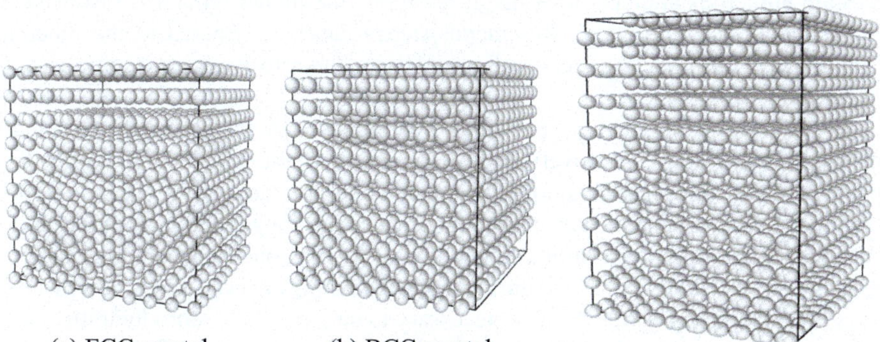

(a) FCC crystal structure (b) BCC crystal structure (c) HCP crystal structure

Fig. 13 Consequence of a generic prompt: failure to detect all valid crystal structures

6.1.2 Short Term Memory

In subsequent iterations, efforts were made to curb the model's random behavior by providing more precise information. To retrieve the equations for primitive and basis vectors from the initial prompt, GPT-4 was specifically instructed to use the primitive vectors from the initial input. However, due to its limited short-term memory, the model encountered difficulties in recalling all the required information from the initial prompt shown in Fig. 10. Consequently, primitive and basis vectors were reintroduced as input prompts to ensure accuracy. It is noteworthy that the challenges observed with GPT-3.5 persist with GPT-4, highlighting the continued need for careful and explicit instructions to overcome limitations in model memory and enhance overall performance.

6.1.3 Code Reproducibility

In the subsequent iteration, the model successfully generated an exact code capable of producing all three crystal structures in arbitrary orientations. Acknowledging the significant influence of prompt engineering on the performance of LLMs, GPT-4 was specifically tasked with generating a prompt that would enable the production of precise code consistently. Despite incorporating a comprehensive prompt as depicted in Fig. 14 that encapsulated all the necessary information for the precise regeneration of the code, the model encountered challenges in reproducing identical results.

Due to the constrained field knowledge and limited thinking capability, the model encountered difficulties in distinguishing between the concepts of the number of unit cells and the calculation of box length considering periodicity. Specifically, it struggled with prompts related to periodicity calculation, particularly with the HCP structure generation, as HCP possesses a distinct lattice constant in the z-direction. Despite numerous attempts to generate prompts, the model remained inconsistent in producing accurate code. This inconsistency can be attributed to the inherent limitations of the model's understanding and its tendency to overlook specific aspects of the input prompts.

In addition, the model's performance was hindered by the abundance of inputs provided, which may have overwhelmed its attention and led to confusion. When presented with an excessive amount of information, the model may struggle to prioritize and comprehend the key components necessary for generating accurate code. As a result, splitting the prompt into two distinct parts—one for calculating the box length with periodicity Fig. 3 and the other for generating a crystal structure with arbitrary orientation Fig. 4 was deemed necessary to enhance code reproducibility. This approach aims to streamline the input process and mitigate the model's tendency to overlook essential details, ultimately improving the reliability of the generated code.

After splitting the entire code into smaller, more manageable prompts, the randomness in the model's responses became controllable. This allowed the model to produce code for calculating periodicity more accurately, but it still required a few iterations to refine the output and get the exact code for structure generation. Overall, this approach proved to be far better compared to using a single prompt for generating the entire code.

Objective:

- Develop a complete Python script to calculate and output crystal structures for Face-Centered Cubic (FCC), Body-Centered Cubic (BCC), and Hexagonal Close Packed (HCP) systems, focusing on precise atom position coordinates generation and saving.

Core Functionality:
Function Name: generate_crystal_structure
Inputs Required:

- Crystal Type: Must be one of 'FCC', 'BCC', or 'HCP'.
- Lattice_Constant 'a': Fundamental for all crystal types.
- Additional Lattice Constant 'c': Required for HCP only, with the condition $c > a$.

- number_of_unit_cells (x,y,z): The number of unit cells in each direction for the simulation box.
- Orientation: Defined by a tuple containing three tuples, each for Miller indices in x, y, and z directions, respectively.

Primitive and Basis Vectors:
These vectors define the initial layout of atoms within a unit cell and it should be rotated based on the given orientation.

- FCC: Primitive vectors [[0.5*a, 0.5*a, 0], [0.5*a, 0, 0.5*a], [0, 0.5*a, 0.5*a]], basis vectors [0, 0, 0].
- BCC: Primitive vectors [[-0.5*a, 0.5*a, 0.5*a], [0.5*a, -0.5*a, 0.5*a], [0.5*a, 0.5*a, -0.5*a]], basis vectors [0, 0, 0].
- HCP: Primitive vectors [[0.5*a, -0.5*(3**(1/2))*a, 0], [0.5*a, 0.5*(3**(1/2))*a, 0], [0, 0, c]], basis vectors [[0.5*a, 0.5*(3**(-1/2))*a, 0.25*c], [0.5*a, -0.5*(3**(-1/2))*a, 0.75*c]].

Rotation Matrix Generation:
Function: rotation_matrix_from_miller_indices(*hkl_x*, *hkl_y*, *hkl_z*)

- Constructs a rotation matrix from Miller indices, essential for orienting the crystal structure.

Formulas to Calculate Periodicity Length:

- For FCC & BCC:

Fig. 14 Final prompt generate by GPT4

  - Periodicity_length = number_of_unit_cells × Lattice_Constant 'a' × norm(primitive_vectors)
- For HCP in x and y direction:
  - Periodicity_length = number_of_unit_cells × Lattice_Constant 'a' × norm(primitive_vectors)
- For z direction in HCP:
  - Periodicity_length = number_of_unit_cells × Lattice_Constant 'c' × norm(primitive_vectors)
- Floor down the output and provide as integers.

Notes:

- Primitive vectors and basis vectors should be rotated before the Atom Position Generation.

Atom Position Generation:

- Important: All positions must be within defined box lengths and checked for validity.

- For Loop Ranges to calculate atom position: The for loop should range between (-Periodicity_length, Periodicity_length, 1) for FCC and BCC, and adjusted ranges for HCP, ensuring complete coverage of the simulation box. Position Checks: Ensure positions are > 0 and within (Periodicity_length + 1) for FCC/BCC, and similarly adjusted for HCP.

Output Specifications:

- File Format: XYZ, named Crystal_structure.xyz, listing all calculated atom positions.

User Interface:

- Prompt for crystal type, lattice constants, number of unit cells, and orientation, ensuring input validation and error handling, especially for orthogonality of orientation vectors and lattice constants validity.

Error Handling:

- Include checks for input validity, such as crystal types, lattice constants, and orthogonality of orientation vectors.

Fig. 14 (continued)

Table 7 List of frequent errors and subsequent prompts

Frequent errors	Subsequent prompts
Validation of atom position	The condition used to check the atom positions is not precisely followed in the code.
For loop ranges	The loop range condition needs to be strictly adhered to as specified in the input prompt.
Rotation	The primitive and basis vectors have to be rotated initially.
Relative distance calculation	x1 and x2 should be calculated as the relative distance between the dislocation line and the atom position.
Position of dislocation	The dislocation position should be taken as input based on the dislocation direction provided by the user.

7 Minor Errors

The code generated by the model might need some adjustments, as there were common errors during its development. Table 7 are the frequent issues identified along with the additional guidance provided for rectification. These errors have been frequently encountered, but the subsequent prompts provided offer resolutions to rectify them effectively. With these modifications implemented, one can replicate the process with minimal issues, albeit with some minor periodicity-related issues.

8 Discussion

Utilizing GPT-4 for evaluating its potential as automated programming assistance across a range of atomistic simulation tasks has generally yielded satisfactory outcomes. However, as the complexity of the tasks escalates, deviations from the user prompts become more apparent in the model's behavior. Here, we present a comprehensive discussion outlining observations on GPT-4's performance in handling intricate tasks.

- **Limitation in Handling Complex Tasks:** The model encountered difficulties in accurately generating all three basic crystal structures without precise user inputs. Challenges arose in properly populating the simulation box and validating atom positions, particularly without explicit instructions such as employing for loops and verifying positions. The HCP crystal structure posed significant hurdles due to its intricate nature, involving distinct lattice parameters ('a' & 'c') and additional basis vectors, in contrast to FCC and BCC. This underscores the model's limitations in handling complex structures. Furthermore, when arbitrary rotations were introduced, although the model could generate the rotation matrix, it struggled to apply it accurately to the primitive vectors and basis vectors. Consequently, human feedback remained essential even with clear and concise prompts, highlighting the ongoing need for human intervention in such scenarios. Despite receiving feedback, issues pertaining to the periodicity of surface atoms persist in the final output shown in Figs. 5 and 6.

- **Autonomous Decision-Making in Mathematical Computations:** However, it is noteworthy that the model's proficiency in handling tasks related to dislocation insertion surpasses its capability in generating single crystals with arbitrary rotations. In this scenario, it autonomously figured out the precise equations shown in (3) and (4) for calculating displacements, even in the absence of explicit user prompts. Furthermore, it opted to utilize the `np.arctan2(y, x)` function from NumPy instead of `np.arctan(y/x)`, as the former better accounts for quadrant distinctions, unlike the latter, which cannot differentiate between quadrants. These decisions were made without specific instructions from the user, highlighting the models ability to autonomously make informed choices in its computations. These capabilities suggest an advanced level of mathematical understanding and problem-solving skills in GPT-4. However, it's important to note that while the model-generated code demonstrates competency in many areas, minor issues may arise still in the insertion of dislocations in specific orientations. Nonetheless, it has consistently showcased its capability to generate entire code segments within a few iterations, a notable feat when compared to single crystal generation.
- **Sensitivity to User Prompts and Memory Constraints:** Moreover, the model's responses are highly sensitive to even minor changes in the input prompt, resulting in inconsistent outputs. Additionally, it has been observed that GPT-4 sometimes struggles to maintain coherence and consistency over extended conversational contexts [7]. Understanding the memory mechanism of Chat-GPT plus is essential for optimizing interactions with the AI. Chat-GPT plus, based on the GPT-4 architecture, has a short-term memory capacity limited by an 8,000-token constraint [6]. As this limit is reached, the AI begins to forget the earliest parts of the conversation, unable to recall details beyond its immediate context window. Instead, it relies on patterns and knowledge acquired during its training on a vast collection of internet texts to generate relevant responses. Recognizing this limitation is crucial for generating effective prompts and achieving more meaningful engagements with GPT-4 [5]. Of course, well-established software systems for Molecular Dynamics (MD) simulations are available with inbuilt atomic structure generation such as, e.g., LAMMPS [9] or Atomsk [3]. In functionalities they clearly go far beyond the code presented in this chapter; from straightforward crystal structures to more intricate designs like nano-wires, and non-periodic structures.

9 Conclusion

In conclusion, the study aimed to evaluate GPT-4's capability in generating standalone code for constructing single crystals with arbitrary orientations, including dislocation insertion. The assessment involved examining the code generated by

GPT-4 in response to user-defined prompts using visualization software OVITO. The findings indicate that, with extensive human feedback, the generated results generally met user requirements satisfactorily. Large Language Models (LLMs) such as GPT-4 have emerged as crucial assets across various industries. However, their effectiveness relies heavily on well-crafted prompts as user input, highlighting the significance of prompt engineering. Crafting effective prompts requires human intervention and expertise in the relevant field to navigate and harness the inherent randomness of these models. Without deep domain understanding, generating precise and specific prompts becomes challenging. Thus, while sophisticated LLMs offer immense potential, human knowledge remains superior, especially in tackling intricate tasks. However, a burgeoning programming language known as prompt engineering is poised to play a pivotal role in the future, aiding humans in their respective domains.

References

1. Anderson, P.M., Hirth, J.P., and Lothe, J. Theory of Dislocations, 2017, Cambridge University Press.
2. Callister, W., Rethwisch, D. Materials Science and Engineering: an Introduction. (John Wiley & Sons, New York, 2007).
3. Hirel, P. Atomsk: A tool for manipulating and converting atomic data files, pp. 212-219 (2015) https://www.sciencedirect.com/science/article/pii/S0010465515002817.
4. Hull, D., and Bacon, D.J. Introduction to dislocations. Vol. 37. Elsevier, 2011.
5. OpenAI ChatGPT memory span. (https://www.4fsh.com/), [Online; accessed Feb-2024]
6. (<i> Chat GPT-4 Plus Memory Limit – Community.Openai.Com</i>, n.d.) [Online; accessed Feb-2024].
7. Ray, P. ChatGPT: A comprehensive review on background, applications, key challenges, bias, ethics, limitations and future scope. *Internet Of Things And Cyber-Physical Systems.* **3** pp. 121-154 (2023), https://www.sciencedirect.com/science/article/pii/S266734522300024X
8. Stukowski, A. Visualization and analysis of atomistic simulation data with OVITO-the Open Visualization Tool. *Modelling And Simulation in Materials Science And Engineering* (2010).
9. Thompson, A.P., Aktulga, H.M., Berger, R., et al. LAMMPS - a flexible simulation tool for particle-based materials modeling at the atomic, meso, and continuum scales, pp.108171 (2022), https://www.sciencedirect.com/science/article/pii/S0010465521002836.

Molecular Dynamics Simulation of Noble Gases

Aagashram Neelakandan, Vishal Vijendra Badami, and Bernhard Eidel

Abstract This chapter presents a detailed exploration of a Python code generated by ChatGPT-4 for the 2D Molecular Dynamics simulation of noble gases, with a specific focus on Argon. The chapter discusses the implementation of the velocity Verlet algorithm for integrating Newton's second law and the application of the Lennard-Jones (12–6) pair potential to model the interactions between Argon atoms. The code is rigorously verified through a series of tests, emphasizing energy conservation within an NVE ensemble. Both simple two-atom systems and more complex multi-atom simulations within a periodic boundary condition framework are analyzed, demonstrating the code's accuracy and reliability in simulating molecular dynamics in noble gases.

1 Introduction

Molecular dynamics (MD) simulations touch on many aspects of physics and are a valuable resource for comparing theoretical models to experimental results. MD uses computer simulation with statistical mechanics to compute static and dynamic properties of a classical many-body system [1–4]. In contrast to Molecular Statics (MS) whose simulations are carried out at a temperature of 0 K, MD simulations are carried out at a temperature greater than 0 K. The classical MD method simply solves numerically Newton's equations of motion for the interacting many-particle with pair potentials. Here, we treat atoms as classical Newtonian particles and we can compute the acceleration of any atom. The force between atoms is still determined

A. Neelakandan (✉) · V. V. Badami · B. Eidel
M5, IMFD, TUBAF, Freiberg, Germany
e-mail: aagashram.neelakandan@student.tu-freiberg.de

V. V. Badami
e-mail: vishal-vijendra.badami@student.tu-freiberg.de

B. Eidel
e-mail: bernhard.eidel@imfd.tu-freiberg.de

© The Author(s), under exclusive license to Springer Nature Switzerland AG 2025

B. Eidel (ed.), *GPT for Python-Coding in Computational Materials Science and Mechanics*, Studies in Computational Intelligence 1198,
https://doi.org/10.1007/978-3-031-85470-5_2

via the interatomic potential by the gradient of the pair potential function. In this project, we are using Lennard-Jones potential (LJ potential) also known as 12–6 potential.

This is a valid potential for chemical inert gases such as Argon (Ar). Argon is a noble gas. It is one of the elements in Group 18 of the periodic table, which includes other noble gases like helium, neon, krypton, xenon, and radon. These gases are characterized by their lack of reactivity due to having a full valence electron shell, making them very stable and unlikely to form chemical bonds under normal conditions. Argon is the third noble gas, following helium and neon.

Temperature and time play a role in MD simulations as compared to MS simulations. There are many ensembles in MD, the NVE (microcanonical) ensemble, NVT (canonical) ensemble, and NPT ensemble. Here variables are, N is the number of particles, V is the volume of the simulation box, E is the total energy of the simulation box, T is the temperature of the simulation box and P is the pressure in the simulation box. In the NVE ensemble, variables N, V, and E are kept constant and this rule applies to all the above ensembles. For NVT and NPT ensembles, additional couplings will be used such as thermostats and barostats respectively.

1.1 Interatomic Potential

At the core of MD simulations are interactions between the individual molecules. These interactions can be separated into two types, bonded and non-bonded. Bonded interactions take place within molecules, between atoms which are connected in some way. Bonded interactions might include terms to change the length of a chemical bond, or change the angle of a bond. All other interactions between atoms are classed as non-bonded interactions. This simulation consists entirely of Argon atoms, so only non-bonded interactions need to be considered. These can be modeled with the LJ potential.

The LJ potential $V(r)$ is given by:

$$V(r) = 4\epsilon\left[\left(\frac{\sigma}{r}\right)^{12} - \left(\frac{\sigma}{r}\right)^{6}\right], \tag{1}$$

where $V(r)$ is the potential energy as a function of the distance r between two particles, ϵ is the depth of the potential well, which represents the strength of the attractive interaction, here the value is 1.65×10^{-21} J, σ is the finite distance at which the inter-particle potential is zero, also known as the collision diameter, here the value is 3.4×10^{-10} m.

The LJ potential consists of a short-range repulsive term and a (relatively) long-range attractive term. The r^{-6} attractive term comes from the London dispersion force, which is (the weakest) part of the Van der Waals forces. Van der Waals forces are weak attractions between atoms that do not cause chemical bonds to be formed. Often called induced-dipole dipole interactions.

The short-range repulsive term models the Pauli exclusion principle by stopping particles from getting too close. The r^{-12} exponent does not have a rigorous physical justification. It is computationally efficient as it is just the square of the other r^{-6} term.

Justification for Use in Modeling Noble Gases Noble gases like argon are characterized by their closed-shell electronic configurations, meaning that they have no permanent dipole moments and are chemically inert. The interactions between noble gas atoms are primarily due to weak van der Waals forces. The Lennard-Jones potential is particularly well-suited to modeling these interactions for several reasons:

- **Van der Waals Forces Dominance**: For noble gases, the interaction is predominantly due to dispersion forces, which the Lennard-Jones potential effectively captures through its $(\sigma/r)^6$ term.
- **Simplicity and Computational Efficiency**: The Lennard-Jones potential is computationally simple and efficient to calculate, making it ideal for large-scale simulations of noble gases where more complex potentials might be unnecessary or impractical.
- **Empirical Fit to Experimental Data**: Parameters ϵ and σ can be fitted to experimental data, allowing the Lennard-Jones potential to accurately reproduce the properties of noble gases such as their phase behavior (liquid-gas coexistence curve), transport properties (viscosity, diffusion), and thermodynamic properties (equation of state).
- **Historical Success**: The Lennard-Jones potential has a long history of successful application in simulating noble gases. It was originally developed based on experimental observations of argon, and it has been validated extensively through its ability to reproduce the experimentally observed behaviors of noble gases.

Limitations and Considerations While the Lennard-Jones potential is highly effective for noble gases, the limitations shall be briefly mentioned. It is the (i) lack of directionality, since LJ potential does not account for any directional dependence in bonding, which is unimportant for noble gases but would be a limitation for systems involving directional bonds (e.g., covalent bonds in molecules). Furthermore, the (ii) approximation of repulsion through the r^{-12} term is an approximation of the repulsive forces, chosen primarily for computational convenience. While it works well for noble gases, more complex repulsive terms might be needed for other systems. As another shortcoming, (iii) the temperature and pressure dependence; LJ potential can not be universally accurate across all temperatures and pressures, since the parameters ϵ and σ are usually fitted for specific conditions.

1.2 Newton's 2nd Law of Motion and Time Integration

Newton's second law of motion is given by the time derivative of the linear momentum $\boldsymbol{p} = m\boldsymbol{v}$ and the force $\boldsymbol{F}$ according to

$$\boldsymbol{F} = \frac{\mathrm{d}\boldsymbol{p}}{\mathrm{d}t} \tag{2}$$

which simplifies for constant masses, hence $\dot{m} = 0$ to

$$\boldsymbol{F} = m\boldsymbol{a}\,, \tag{3}$$

where m is the mass of the particle and $\boldsymbol{a}$ its acceleration.

The positions and velocities of the particles are evaluated using the Velocity Verlet integration method, where the predicted positions $\boldsymbol{x}$, predicted velocities $\boldsymbol{v}$, and predicted acceleration $\boldsymbol{a}$ at time $t + \Delta t$ are obtained from the same, given quantities at the current time t in the following way:

$$\boldsymbol{x}(t + \Delta t) = \boldsymbol{x}(t) + \boldsymbol{v}(t)\Delta t + \frac{1}{2}\boldsymbol{a}(t)\Delta t^2\,, \tag{4}$$

$$\boldsymbol{v}(t + \Delta t) = \boldsymbol{v}(t) + \frac{\boldsymbol{a}(t) + \boldsymbol{a}(t + \Delta t)}{2}\Delta t\,, \tag{5}$$

where Δt is the time step size over which the integration is performed to obtain the predicted quantities. Velocity Verlet time integration method is employed since it is fast, requires little memory, and is easy to use for long time steps.

In this work, boundary conditions (BC) are periodic. They mimic the behavior of the infinite bulk surrounding the sample. In this way, surface effects are removed. It should also be noted that if the particles go through a boundary of the simulation box, they will appear on the other side of the box.

We will present an MD simulation of Argon particles for the gas phase, which is considered to be moving with the velocity given by the Maxwell-Boltzmann distribution as shown in Fig. 1. The Maxwell-Boltzmann distribution describes the velocity of particles in ideal gases at thermodynamic equilibrium, where particles exchange energy through brief collisions and their velocities follow Maxwell-Boltzmann statistics based on kinetic energy. In MD simulations we first specify the initial positions and momentum of the particles, the latter refers to the Maxwell-Boltzmann distribution.

1.3 Statistical Ensemble

Statistical ensembles define the conditions under which atomistic simulations are carried out, dictating what thermodynamic variables are conserved or controlled. Different ensembles correspond to different sets of thermodynamic variables that are held constant during the simulation. Below is a description of the most common ensembles of the microcanonical ensemble NVE, the canonical ensemble NVT, and the isothermal-isobaric ensemble NPT, where the acronyms reflect the variables

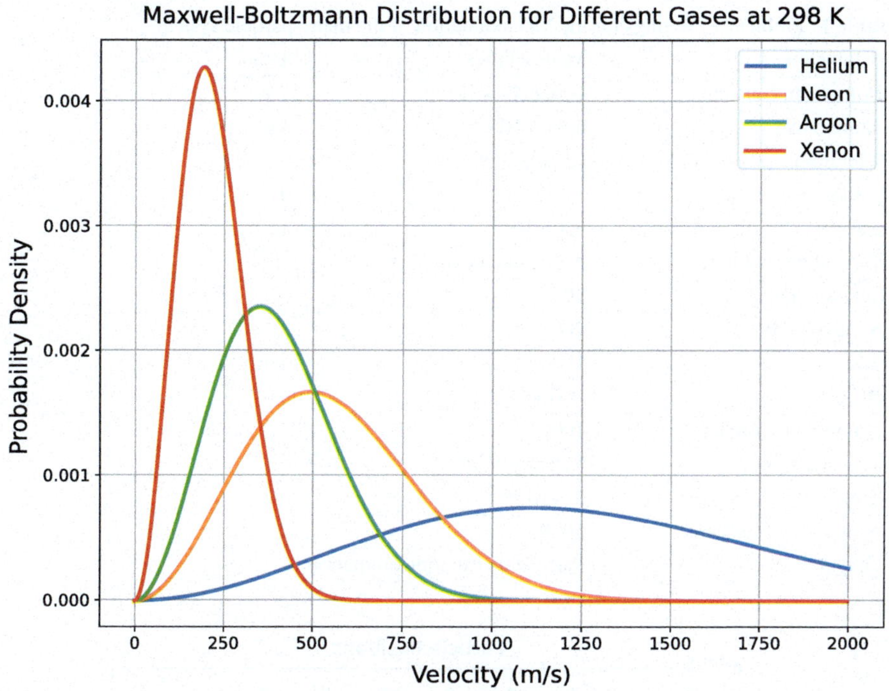

Fig. 1 Probability density functions of Maxwell-Boltzmann distribution for the velocities of noble gases like Helium, Neon, Argon and Xeon at a temperature of 298 K

involved, the number of particles N, the particle volume V, the energy E, the temperature T, and the pressure P.

Here we use the NVE ensemble which keeps constant N, V and the internal energy E which follows from the system characteristic being isolated. It meets the requirements of the present goals simulating systems with no heat exchange with the surroundings, analogous to an isolated system in thermodynamics. It provides a natural way to observe the dynamical evolution of a system without external influences, often used for studying the intrinsic properties of a system.

2 Prompt

The parameters and constants relevant to the simulation are given in Table 1.

Figure 2 shows the initial positions of the particles for the simulation. Then we evolve the system according to Newton's second law of motion for which we let the particles interact through a LJ potential using an NVE ensemble. Finally, we measure physical quantities as functions of particle positions and momentum.

Table 1 Problem 1: Settings for the MD simulation of the many-particle system

Parameters	Values/Types	Units
Boltzmann constant k_B	1.380649×10^{-23}	J/K
Atomic mass (Ar)	6.63×10^{-26}	kg
Interatomic potential (Ar)	LJ potential	
with ϵ	1.65×10^{-21}	J
and σ	3.4×10^{-10}	m
Type of ensemble	NVE (microcanonical)	
Temperature T	300	K
Simulation box	2D	
Size	10×10	nm
BC	Periodic	
Number of particles	100	
Time integrator	Velocity Verlet	
Time step size Δt	2×10^{-15}	s
Number of time steps	5000	
Output	Particle distribution animation	

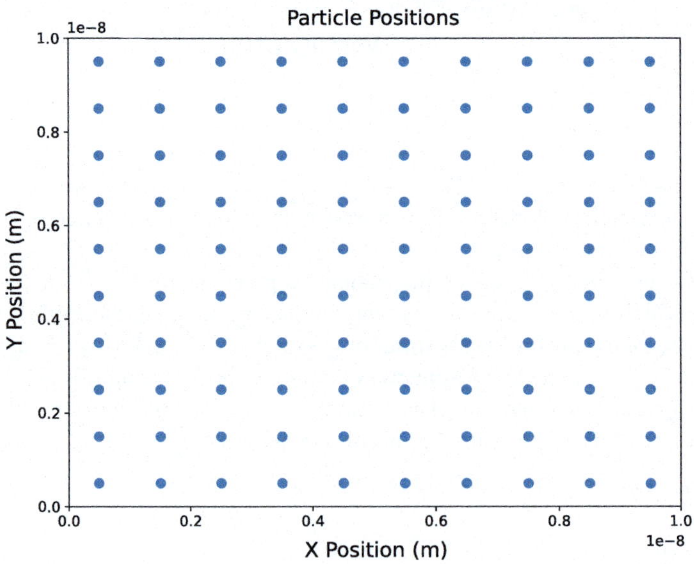

Fig. 2 Initial particle positions in the 2D MD simulation for argon gas

The algorithm for the 2D MD simulation is given as pseudocode in the algorithm box 1

Algorithm 1: Molecular Dynamics Simulation for 2D System

Input: Number of particles N, initial positions $\boldsymbol{x}_i(0)$, initial velocities $\boldsymbol{v}_i(0)$, time step Δt, number of time steps N_p, force field parameters

Output: Particle trajectories $\boldsymbol{x}_i(t)$, velocities $\boldsymbol{v}_i(t)$, and plot kinetic energy $E_{\text{kin}}(t)$, potential energy $E_{\text{pot}}(t)$, and total energy $E_{\text{tot}}(t)$ over time step Δt

Initialization:

Initialize the velocities of the particles such that they follow a Maxwell-Boltzmann distribution corresponding to a temperature of 300 K. Ensure the system has zero net momentum by adjusting the velocities;

Set initial positions $\boldsymbol{x}_i(0)$ and velocities $\boldsymbol{v}_i(0)$ for all particles $i = 1, \ldots, N$;

Compute initial forces $\boldsymbol{F}_i(0)$ on each particle using the interatomic potential;

for $t = 0$**to**T *with step* Δt **do**

  Velocity Verlet Integration:

  foreach *particle* i **do**

    Update positions:

$$\boldsymbol{x}_i(t + \Delta t) = \boldsymbol{x}_i(t) + \boldsymbol{v}_i(t)\Delta t + \frac{\boldsymbol{F}_i(t)}{m_i}\frac{\Delta t^2}{2}$$

    Compute intermediate velocities:

$$\boldsymbol{v}_i(t + \frac{\Delta t}{2}) = \boldsymbol{v}_i(t) + \frac{\boldsymbol{F}_i(t)}{m_i}\frac{\Delta t}{2}$$

  end

  Force Calculation:

  Recalculate forces $\boldsymbol{F}_i(t + \Delta t)$ on each particle due to interactions using the interatomic potential;

  foreach *particle* i **do**

    Update velocities:

$$\boldsymbol{v}_i(t + \Delta t) = \boldsymbol{v}_i(t + \frac{\Delta t}{2}) + \frac{\boldsymbol{F}_i(t + \Delta t)}{m_i}\frac{\Delta t}{2}$$

  end

  Apply Boundary Conditions:

  Apply boundary conditions (e.g., periodic boundaries, reflective walls) to updated positions and velocities. Here periodic boundary conditions are to be applied;

  Calculate System Properties:

  Compute kinetic energy $E_{\text{kin}}(t)$, potential energy $E_{\text{pot}}(t)$, and total energy $E_{\text{tot}}(t)$;

end

Output:

Return particle trajectories $\boldsymbol{x}_i(t)$, velocities $\boldsymbol{v}_i(t)$, and plot $E_{\text{kin}}(t)$, $E_{\text{pot}}(t)$, and $E_{\text{tot}}(t)$ over time step Δt;

The final prompt which worked is given in Fig. 3. It was used for generating the Code Listing 2.1 is,

Generate a Python code for the molecular dynamics simulation for argon gas in a 2D rectangular domain. The simulation box has dimensions of 10 nanometers by 10 nanometers. Consider the FCC lattice and make sure to initialize the positions of atoms in the FCC lattice itself to avoid the overlapping of the atoms.

Initialization: Initialize the positions of 100 particles randomly within the simulation box ensuring that no two particles overlap. Initialize the velocities of the particles such that they follow a Maxwell-Boltzmann distribution corresponding to a temperature of 300 Kelvin. Ensure the system has zero net momentum by adjusting the velocities.

Integration Method: Use the robust Velocity Verlet integration method to update the positions and velocities of particles based on the forces acting on them.

Periodic Boundary Conditions: Apply periodic boundary conditions directly after updating the positions to maintain the particles within the simulation box.

Interatomic Potential: Use the Lennard-Jones potential for calculating interatomic forces, with an appropriate small epsilon value to avoid singularities in the denominator.

Simulation Parameters: Set the time step for the simulation to 2 femtoseconds and simulate 5000 time steps.

Ensemble and Conservation: Use the NVE (microcanonical) ensemble. As atoms interact with each other stored potential energy is converted to kinetic energy and keep in mind since it's an NVE ensemble the total energy of the system should remain constant.

Output: Plot the kinetic energy, potential energy, and total energy to the number of time steps NOT Time.
Print the values of kinetic energy, potential energy and total energy at the end of the simulation.
Create a snapshot in the form of an image to visualize particles and save the resulting snapshot output in the user's desktop path.

Create separate functions for: Initializing positions and velocities of the

Fig. 3 Prompt for generating a 2d MD code in python for argon gas

particles.
Computing forces and potential energy.
Velocity Verlet integration.

The main simulation loop.
Create the INITIAL and FINAL scatter plots of the particles in the system and save the snapshot using the below code snippet. PLEASE ENSURE THAT PARTICLES SHOULD NOT BE ON THE EDGE OF THE BOX DIMENSIONS.

```
def get_desktop_path():
    home = os.path.expanduser("~") # Gets the home
    ↪ directory
    desktop = os.path.join(home, 'Desktop')  # Appends
    ↪ Desktop to the home path
    return desktop
```

Ensure all variables are correctly defined and avoid typographical errors in the code. Use precise and accurate scaling factors and conversion factors for all physical quantities. Make sure the lengths of the time array and the energies array are correct.

Also, there are two tests to be conducted for verification:
Test 1: Test for periodicity, Consider one atom, which flies through the simulation box. Check, whether it reenters the simulation box from the left, once it has passed through the right bounding box. The output should be in the form of an animation-like GIF file using Pillowriter and should be saved in the user's desktop folder with all the proper labeling on both axes.
Test 2: Quantitative Test for Energy Conservation. In the simulation box there are only two atoms, which fly with the same speed on the same line but against each other for a perfect, elastic hit - and then fly exactly backward. In that case, you can easily calculate the kinetic energy of the atoms. (Doing so you can verify the value from the simulation.) A diagram with the kinetic energy over time leads to a drop to zero at the hit when velocity is zero for both of them - and the potential energy takes its maximum. Of course, this requires a manipulation of the initial conditions of the simulation. The output should be in the form of an animation-like GIF file using Pillowriter and should be saved in the user's desktop folder with all the proper labeling on both axes.

Avoid errors like:
operands could not be broadcast together with shapes (64,2) (100,2).
operands could not be broadcast together with shapes (25,2) (100,2).
operands could not be broadcast together with shapes (50,2) (100,2).
Undefined variables.

Fig. 3 (continued)

typographical errors in the code.
index 50 is out of bounds for axis 0 with size 50.
overflow encountered in a square.
division by zero and invalid values arising from the Lennard-Jones potential calculation.
make sure to properly handle to avoid the divide-by-zero error and ensure no invalid values are encountered.
As atoms interact with each other, stored potential energy is converted to kinetic energy. Keep in mind that since it's an NVE ensemble, the total energy of the system should remain constant.

Fig. 3 (continued)

3 Code Listing

The code listing 2.1 displays the python code generated by ChatGPT-4o for the prompt given in Fig. 3.

```
import os
import numpy as np
import matplotlib.pyplot as plt
import matplotlib.animation as animation
from matplotlib.animation import PillowWriter

def get_desktop_path():
    home = os.path.expanduser("~")  # Gets the home
    ↪ directory
    desktop = os.path.join(home, 'Desktop')  # Appends
    ↪ Desktop to the home path
    return desktop

# Constants
kB = 1.38064852e-23  # Boltzmann constant (J/K)
epsilon = 1.65e-21  # Lennard-Jones potential well depth (J)
sigma = 3.4e-10  # Lennard-Jones potential distance
    ↪ parameter (m)
mass = 6.63e-26  # Mass of argon atom (kg)
T = 300  # Temperature (K)
L = 10e-9  # Length of the simulation box (m)
num_particles = 100  # Number of particles
dt = 2e-15  # Time step (s)
num_steps = 5000  # Number of simulation steps

def initialize_positions(num_particles, L):
    """Initialize positions of particles in an FCC lattice
    ↪ """
    positions = np.zeros((num_particles, 2))
    num_cells_per_side = int(np.ceil(np.sqrt(num_particles))
    ↪ )
    cell_size = L / num_cells_per_side
```

```
    particle_idx = 0

    for i in range(num_cells_per_side):
        for j in range(num_cells_per_side):
            if particle_idx >= num_particles:
                break
            positions[particle_idx] = np.array([i * cell_size, j * cell_size])
            particle_idx += 1

    # Ensure particles are not on the edge of the box
    positions += cell_size / 2
    return positions

def initialize_velocities(num_particles, T, mass):
    """Initialize velocities of particles to follow Maxwell-Boltzmann distribution"""
    velocities = np.random.normal(0, np.sqrt(kB * T / mass), (num_particles, 2))
    velocities -= np.mean(velocities, axis=0)  # Ensure zero net momentum
    return velocities

def compute_forces(positions):
    """Compute forces and potential energy using Lennard-Jones potential"""
    forces = np.zeros_like(positions)
    potential_energy = 0.0

    for i in range(len(positions)):
        for j in range(i + 1, len(positions)):
            r_ij = positions[i] - positions[j]
            r_ij -= L * np.round(r_ij / L)  # Apply periodic boundary conditions
            r2 = np.dot(r_ij, r_ij)
            if r2 < (3 * sigma) ** 2:  # Apply a cutoff to save computation
                r2_inv = sigma ** 2 / r2
                r6_inv = r2_inv ** 3
                r12_inv = r6_inv ** 2
                force_magnitude = 24 * epsilon * (2 * r12_inv - r6_inv) / r2
                forces[i] += force_magnitude * r_ij
                forces[j] -= force_magnitude * r_ij
                potential_energy += 4 * epsilon * (r12_inv - r6_inv)

    return forces, potential_energy

def velocity_verlet(positions, velocities, forces, dt):
    """Perform a single step of Velocity Verlet integration"""
    positions += velocities * dt + 0.5 * forces * dt ** 2 / mass
```

```
    positions = positions % L  # Apply periodic boundary
    ↪ conditions
    new_forces, potential_energy = compute_forces(positions)
    velocities += 0.5 * (forces + new_forces) * dt / mass
    return positions, velocities, new_forces,
    ↪ potential_energy

def run_simulation(num_particles, L, T, mass, dt, num_steps)
    ↪ :
    positions = initialize_positions(num_particles, L)
    velocities = initialize_velocities(num_particles, T,
    ↪ mass)
    forces, potential_energy = compute_forces(positions)
    kinetic_energy = 0.5 * mass * np.sum(velocities ** 2)

    kinetic_energies = []
    potential_energies = []
    total_energies = []

    for step in range(num_steps):
        positions, velocities, forces, potential_energy =
    ↪ velocity_verlet(positions, velocities, forces, dt)
        kinetic_energy = 0.5 * mass * np.sum(velocities **
    ↪ 2)
        total_energy = kinetic_energy + potential_energy

        kinetic_energies.append(kinetic_energy)
        potential_energies.append(potential_energy)
        total_energies.append(total_energy)

    return positions, kinetic_energies, potential_energies,
    ↪ total_energies

def plot_energies(kinetic_energies, potential_energies,
    ↪ total_energies, filename):
    plt.figure()
    plt.plot(kinetic_energies, label='Kinetic Energy')
    plt.plot(potential_energies, label='Potential Energy')
    plt.plot(total_energies, label='Total Energy')
    plt.xlabel('Time Step')
    plt.ylabel('Energy (J)')
    plt.legend()
    plt.savefig(filename)
    plt.show()

def save_snapshot(positions, filename):
    plt.figure()
    plt.scatter(positions[:, 0], positions[:, 1])
    plt.xlim(0, L)
    plt.ylim(0, L)
    plt.xlabel('X Position (m)')
    plt.ylabel('Y Position (m)')
    plt.title('Particle Positions')
    plt.savefig(filename)
    plt.show()
```

```python

# Test 1: Periodicity Test with Animation
def test_periodicity(L):
    """Test for periodicity by checking if a single atom reenters the box with animation"""
    positions = np.array([[0, L/2]])  # Start the particle at the left edge
    velocities = np.array([[1e4, 0]])  # Increased velocity for faster movement
    fig, ax = plt.subplots()
    ax.set_xlim(0, L)
    ax.set_ylim(0, L)
    ax.set_xlabel('X Position (m)')
    ax.set_ylabel('Y Position (m)')
    particle, = ax.plot([], [], 'ro', markersize=12)

    def init():
        particle.set_data([], [])
        return particle,

    def update(frame):
        nonlocal positions, velocities
        forces = np.zeros_like(positions)  # No other particles, so no forces
        positions, velocities, _, _ = velocity_verlet(positions, velocities, forces, dt)
        particle.set_data([positions[0, 0]], [positions[0, 1]])  # Pass as lists
        return particle,

    num_frames = int(L / (velocities[0, 0] * dt)) + 1  # Ensure enough frames to cover the entire path
    ani = animation.FuncAnimation(fig, update, frames=num_frames, init_func=init, blit=True)
    gif_path = os.path.join(get_desktop_path(), 'periodicity_test.gif')
    ani.save(gif_path, writer=PillowWriter(fps=30))
    plt.close(fig)
    print(f"Periodicity Test animation saved as {gif_path}")

# Test 2: Energy Conservation Test with Animation
def test_energy_conservation():
    """Test for energy conservation with two particles colliding elastically with animation"""
    v_initial = 1e5  # Initial velocity of the atoms (m/s)

    # Initial conditions
    x1 = L / 4  # Initial position of atom 1 (m)
    x2 = 3 * L / 4  # Initial position of atom 2 (m)
    v1 = v_initial  # Initial velocity of atom 1 (m/s)
    v2 = -v_initial  # Initial velocity of atom 2 (m/s)

    # Lists to store positions and energies
    positions1 = []
    positions2 = []
```

```
    kinetic_energies = []

    # Simulation loop
    for step in range(num_steps):
        # Update positions
        x1 += v1 * dt
        x2 += v2 * dt

        # Check for collision and update velocities
        if x1 >= x2:
            v1, v2 = v2, v1

        # Save positions and energies
        positions1.append(x1)
        positions2.append(x2)
        kinetic_energy = 0.5 * mass * (v1**2 + v2**2)
        kinetic_energies.append(kinetic_energy)

        # Reflect atoms at the boundaries (elastic collision
 ↪  with the wall)
        if x1 < 0 or x1 > L:
            v1 = -v1
        if x2 < 0 or x2 > L:
            v2 = -v2

    # Create animation
    fig, ax = plt.subplots()
    ax.set_xlim(0, L)
    ax.set_ylim(0, L)
    ax.set_xlabel('X Position (m)')
    ax.set_ylabel('Y Position (m)')

    line1, = ax.plot([], [], 'ro', label='Atom 1')
    line2, = ax.plot([], [], 'bo', label='Atom 2')

    def init():
        line1.set_data([], [])
        line2.set_data([], [])
        return line1, line2

    def update(frame):
        # Update with sequences instead of single values
        line1.set_data([positions1[frame]], [L / 2])
        line2.set_data([positions2[frame]], [L / 2])
        return line1, line2

    ani = animation.FuncAnimation(fig, update, frames=
 ↪  num_steps, init_func=init, blit=True)

    # Save the animation as a GIF using PillowWriter
    desktop_path = os.path.join(os.path.expanduser("~"), "
 ↪  Desktop")
    gif_path = os.path.join(desktop_path, "
 ↪  atomic_collision_simulation.gif")
    ani.save(gif_path, writer='pillow', fps=60)
```

```
    plt.show()

    print(f"Simulation complete. GIF saved to: {gif_path}")

# Main simulation loop
positions = initialize_positions(num_particles, L)
velocities = initialize_velocities(num_particles, T, mass)
forces, potential_energy = compute_forces(positions)

# Save initial snapshot
save_snapshot(positions, os.path.join(get_desktop_path(), '
    ↪ initial_snapshot.png'))

kinetic_energies = []
potential_energies = []
total_energies = []

for step in range(num_steps):
    positions, velocities, forces, potential_energy =
    ↪ velocity_verlet(positions, velocities, forces, dt)
    kinetic_energy = 0.5 * mass * np.sum(velocities ** 2)
    total_energy = kinetic_energy + potential_energy

    kinetic_energies.append(kinetic_energy)
    potential_energies.append(potential_energy)
    total_energies.append(total_energy)

# Plot energies
plot_energies(kinetic_energies, potential_energies,
    ↪ total_energies, os.path.join(get_desktop_path(), '
    ↪ energy_plot.png'))

# Save final snapshot
save_snapshot(positions, os.path.join(get_desktop_path(), '
    ↪ final_snapshot.png'))

# Print final energies
print(f"Final Kinetic Energy: {kinetic_energies[-1]}")
print(f"Final Potential Energy: {potential_energies[-1]}")
print(f"Final Total Energy: {total_energies[-1]}")

# Run the additional tests with animations
test_periodicity(L)
test_energy_conservation()
```

Code Listing 2.1 Code generated by ChatGPT-4o for solving 2D MD simulation of argon gas

The output of the ChatGPT-4o does not guarantee that the code generated will work without any flaws. Thorough tests are indispensible for that reason.

4 Tests for Verification

In the following, two tests of rather basic nature are carried out. They are based on a one-atom system to check the periodic BC, and a two-atom system which checks quantitatively the energy conservation.

4.1 Test for Periodic BC

Figure 4 shows the trail of one atom moving from the left to the right (snapshots at constant time increments Δt) of the left simulation box; once it leaves that box through its periodic boundary at the right edge, it simultaneously re-enters the simulation box as displayed in the 2nd window. The plot at the bottom of Fig. 4 displays the corresponding kinetic energy for the entire process; the constant value over the full trail of a length twice the box size verifies (kinetic) energy conservation and, implicitly, the velocity being constant.

In conclusion, the test verifies the proper functionality of the periodic boundary condition for the MD simulation.

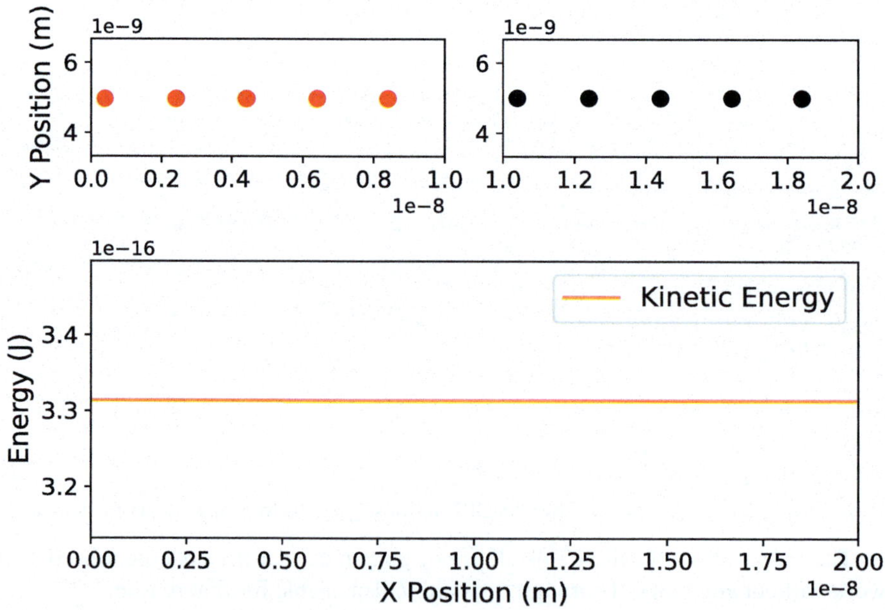

Fig. 4 Periodic BC test for an MD simulation shows the constant kinetic energy as the atom moves across the periodic boundary, from left to right

4.2 Two-Atom-Collision—Interplay of Potential and Kinetic Energies

Figure 5 shows the setup for the energy conservation test. In the simulation box with the size of $L = 10$ nm there are only two atoms (initial positions at $t_0 = 0$: $x_1(t_0) = L/6$, $x_2(t_0) = 5L/6$ with $|x_2(t_0) - x_1(t_0)| \geq r_{\text{cutoff}}$), which fly with the same speed on the same line against each other thus having velocity vectors with opposite signs, (initial velocities $\boldsymbol{v}_1(t_0) = -\boldsymbol{v}_2(t_0)$, $|\boldsymbol{v}_1(t_0)| = 10^4$ m/s), kinetic energy will be maximum in left and right image of Fig. 5, because two particles will be moving with some speed due to attraction and repulsion between them. Whereas potential energy is maximum in center image of Fig. 5. However, in order to make

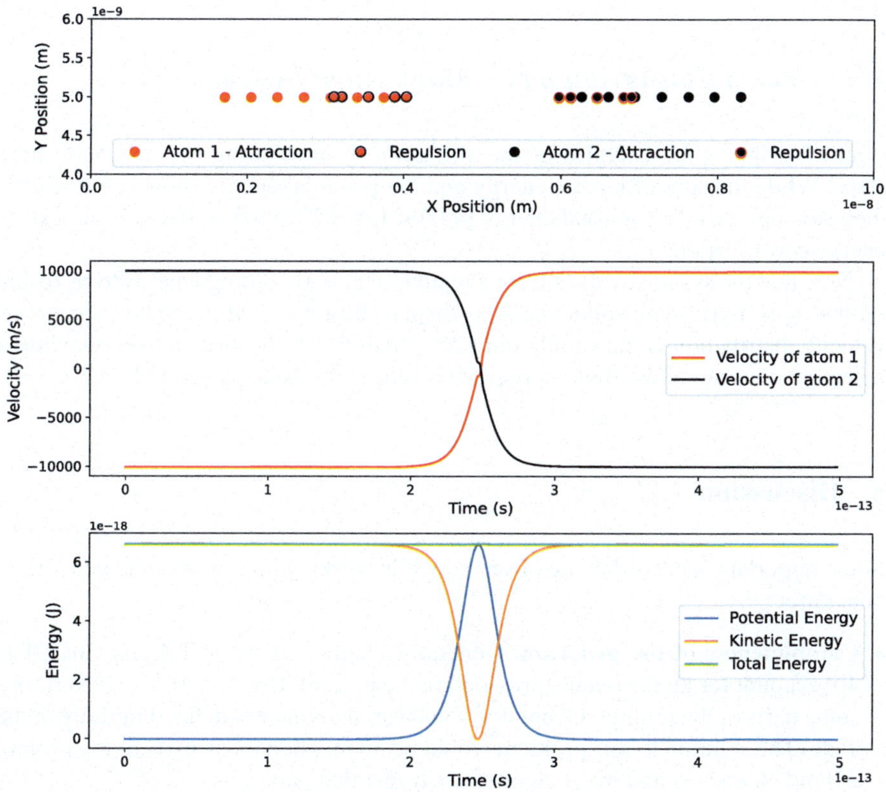

Fig. 5 Quantitative test for energy conservation consisting of two atoms traveling towards each other in the same path with an initial speed of 10^4 m/s. When they come closer, atoms enter the cut-off radius limits of the LJ potential resulting in an increase of the potential energy and decrease in kinetic energy because of their repulsion. This makes the atoms to move in a way from each other resulting in decrease in potential energy and increase in kinetic energy, till they experience the same in when they are near to boundary because of periodic boundary condition

this effect happen within the compatible range of $\Delta t = 2 \times 10^{-15}$ s, parameter σ in the LJ-potential has been increased by one order to 3.4×10^{-9} compared to the argon value tabulated in Table 1.

For this setting, we test the proper calculation of initial energies as well; for $|\boldsymbol{x}_2(t_0) - \boldsymbol{x}_1(t_0)| \geq r_{\text{cutoff}}$, the potential energy at t_0 is zero, $E_{\text{pot}}(t_0) = 0$. The initial kinetic energy is calculated with $v_0 = |\boldsymbol{v}_1(t_0)| = |\boldsymbol{v}_2(t_0)| = 10^4$ m/s and the mass of argon atoms (see Table 1) according to

$$E_{\text{kin}}^{\text{calc}}(t_0) = 2 \cdot \frac{1}{2} m\, v_0^2 = 6.6299999999999994 \times 10^{-18}\text{J} = E_{\text{tot}}(t_0)\,. \tag{6}$$

The figure of $E_{\text{kin}}^{\text{calc}}(t_0)$ from manual calculation exhibits a minor deviation from the value computed by the code $E_{\text{kin}}(t_0) = 6.630605061255307 \times 10^{-18}\text{J}$.

4.3 *Energy Conservation in a Many-Atom System*

Figure 6 displays the energies of the 100-atom system during the first 5000 time steps. While the sum of kinetic energy and the potential energy show some scatter, they sum up to a constant total energy ($4.0592186387714405 \times 10^{-19}$ J) indicating energy conservation.

Note that the system evolves from a regular geometric setting where atoms reside on nodes of a grid with square cells as displayed in Fig. 2. It is the heterogeneous velocity distribution at the simulation start which drives the system to evolve into a heterogeneous atom distribution after 5000 time steps as displayed in Fig. 7.

5 Discussion

Some important aspects that are observed while working on prompts and generating codes are:

- **Completeness of the program**: Generally, chatbots (ChatGPT-4 and ChatGPT-4o) account for all the details provided in the prompt. But sometimes it overlooks some parts of the prompt. Generally, some variables are not defined initially in the code. The solution to this problem would be, to mention those parts of the prompt in bold characters and insist chatbot not forget that part.
- **Short-term memory capacity**: Generally recent chatbots like ChatGPT-4o tend to have long-term memory since they produce the code output as expected for a long time. But after a certain point, chatbots tend to miss some parts of the code and generate the output which is not even mentioned in the prompt. So, we have to write a prompt very specific about the equations to be used, constants, initial variables, time integration method, and output plots.

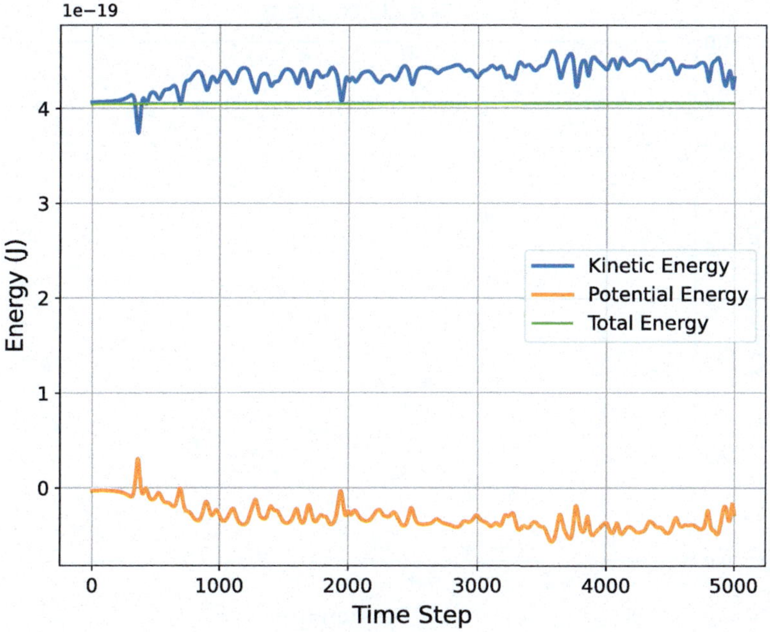

Fig. 6 Energies of the 100-atom system versus the number of time steps indicates the conservation of the total energy while the kinetic and the potential energies show fluctuations

- **Reproducibility**: If we insert the same prompt from the report to the chatbots, the output code will probably be similar but not the same as given in this report. Because chatbots are not consistent with their results. The basic algorithm of the code will be the same, but some minor changes will be there. The recent versions of chatbots like ChatGPT-4o will generate code very similar to those given in this report.
- **Reliability**: The final code generated by the chatbots is reliable with its results only after optimizing the prompts. The codes generated by the recent versions like GPT-4o will be more reliable as compared to the older versions like ChatGPT-4.
- **Hallucinations**: Hallucinations in chatbots refer to instances where the chatbots generate responses that are factually incorrect, nonsensical, or completely fabricated, despite being delivered with apparent confidence. In some cases, We have encountered some minor hallucinations, which can be easily rectified. So, We would say minor or low-level hallucinations occur in chatbots.
- **Learned Lessons**: The main lesson that we have learned is, that the prompt should be very optimized if we need exact and consistent results that match with the research articles. So even minor information must be added to the prompt so that it works perfectly without any illogical errors. The quality and correctness of the output depend mainly on the prompt. So, the prompt should be particular, refined, and optimized.

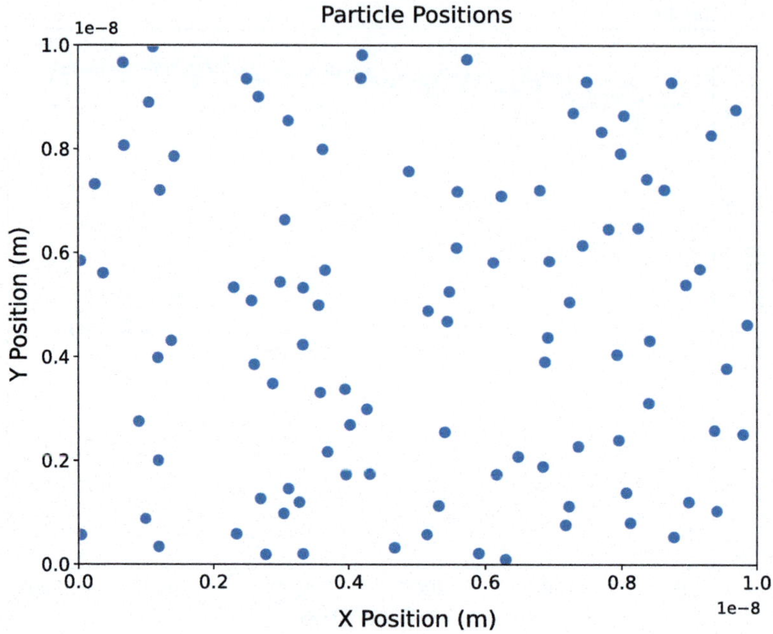

Fig. 7 Heterogeneous atom distribution after time step 5000 in the MD simulation of argon gas

6 Conclusion

The main aim of this chapter was to construct a prompt for ChatGPT-4o to generate a Python code for the 2D MD simulation of noble gases such as argon. To put things into perspective, building blocks of MD simulations were presented such as Newton's equation of motion, the Lennard-Jones (12–6) pair potential, statistical ensembles, and the velocity Verlet algorithm for integrating Newton's second law. The code structure was illustrated by a descriptive pseudocode. The resultant Python code was successfully assessed through a series of tests; the functionality of periodic boundary conditions was verified and for the collision of two atoms the conservation of energy along with its dynamic decomposition into kinetic and potential parts was observed. In conclusion, with some minor manual intervention, ChatGPT-4o has demonstrated its ability to generate an accurate and reliable code for the molecular dynamics simulations of noble gases.

References

1. D. Frenkel, B. Smit, "Understanding molecular simulation: from algorithms to applications", Elsevier, 2023. https://doi.org/10.1016/B978-0-12-267351-1.X5000-7.
2. E.B. Tadmor, R.E. Miller, "Modeling materials: continuum, atomistic and multiscale techniques", Cambridge University Press, 2011. https://doi.org/10.1017/CBO9781139003582.
3. J.M. Haile, "Molecular dynamics simulation: elementary methods", John Wiley & Sons, Inc., 1992. https://doi.org/10.1016/0166-1280(93)87060-q.
4. R. LeSar, "Introduction to computational materials science: fundamentals to applications", Cambridge University Press, 2013. https://doi.org/10.1017/cbo9781139033398.

Phase Field Modeling of Grain Growth

Rahul Narkhede and Bernhard Eidel

Abstract In the current chapter the assignment for GPT-4 on ChatGPT Plus is to generate a Python code for grain growth simulation by the phase field method. Specifically, the non-conserved Allen-Cahn equation with a suitable free energy functional is solved in 2D using the finite difference method and the explicit Euler forward time-stepping scheme. By virtue of a specific prompt design for GPT-4 the resulting Python code allows solving the phase field equations for any generalized initialization. To this end, two variants of initial grain structures are considered; (i) a spherical grain embedded in a larger grain, and (ii) a Voronoi tessellation-based initial structure. Key aspects such as prompt design, code verification and testing of the outcome are discussed.

1 Introduction

Grains of polycrystalline solids constitute the fundamental influencing factors in for most physical properties such as corrosion resistance, thermal and electrical conductivity, and mainly mechanical properties such as strength, ductility and toughness. Usually, the mechanical properties depend on the mean grain size and the grain size distribution. Thus, studying grain evolution becomes a key part of computational materials science.

The local energy at the interface of grains, or grain boundaries, is higher than the corresponding energy in the bulk of the grain. This extra energy at the interface proves the thermodynamic driving force for moving the grain boundary in order to minimize the total free energy. This movement of the grain boundary leads to growth of grains which implies shrinking of others.

R. Narkhede (✉) · B. Eidel
M5, IMFD, TUBAF, Freiberg, Germany
e-mail: rahul-vishnu.narkhede@student.tu-freiberg.de

B. Eidel
e-mail: bernhard.eidel@imfd.tu-freiberg.de

© The Author(s), under exclusive license to Springer Nature Switzerland AG 2025
B. Eidel (ed.), *GPT for Python-Coding in Computational Materials Science and Mechanics*, Studies in Computational Intelligence 1198,
https://doi.org/10.1007/978-3-031-85470-5_3

At the mesoscale, different computational models, such as Monte Carlo Potts model [1], Surface Evolver [2], front-tracking method [3], and cellular automata [4], have been applied. As the number of grains increases, the computational costs become quite demanding due to the need to track individual grain boundaries and apply specific constitutive relations to their evolution. Therefore, phase field (PF) modeling is a potential approach to reduce this computational cost by the description of interfaces as continuous instead of sharp. Several PF models have been proposed for grain growth kinetics. Here we present the use of PF modeling for grain growth simulation based on the example from [5].

In this study, we incorporate the grain growth model of Fan and Chen [6]. In this model, each grain is described by an order parameter η_i, which takes the value of one for a designated grain and zero for all other grains. The evolution of the order parameters is described by the non-conserved Allen-Cahn equation in the form of

$$\frac{\partial \eta_i}{\partial t} = -L_i \frac{\delta F}{\delta \eta_i} \tag{1}$$

for $i = 1, 2, \ldots, N$ grains, where L_i is the mobility coefficient, δ the variational symbol, and F the free energy functional given by

$$F = \int_V \left[f\left(\eta_1, \eta_2, \ldots \eta_N\right) + \sum_i^N \frac{\kappa_i}{2} \left|\nabla \eta_i\right|^2 \right] dv\,, \tag{2}$$

where f is the local free energy density and κ_i are the gradient energy coefficients. The specific form of the orientation-independent f is given in the Fan and Chen model [6] as,

$$f\left(\eta_1, \eta_2, \ldots \eta_N\right) = \sum_i^N \left(-\frac{A}{2}\eta_i^2 + \frac{B}{4}\eta_i^4 \right) + \sum_i^N \sum_{i \neq j}^N \eta_i^2 \eta_j^2\,, \tag{3}$$

in which A and B are positive constants.

The evolution equation governing the numerical implementation is

$$\frac{\partial \eta_i}{\partial t} = -L_i \Big(- A\eta_i + B\eta_i^3 + 2\eta_i \sum_{i \neq j}^N \eta_j^2 - \kappa_i \nabla^2 \eta_i \Big)\,. \tag{4}$$

The Laplacian operator in (4) is approximated using the five-point stencil of the finite difference (FD) method given by,

$$(\nabla^2 \eta)_{i,j} = \frac{\eta_{i+1,j} + \eta_{i-1,j} + \eta_{i,j+1} + \eta_{i,j-1} - 4\eta_{i,j}}{h^2} \tag{5}$$

for a grid spacing $h = h_x = h_y$. The time integration is carried out using the explicit Euler time stepping scheme and thus, we obtain the following discretized expression

$$\frac{\partial \eta_i}{\partial t} = \frac{\eta_i^{n+1} - \eta_i^n}{\Delta t} = -L\Big(- A\eta_i^n + B(\eta_i^n)^3 + 2\eta_i^n \sum_{i \neq j}^{N} (\eta_i^n)^2 - \kappa_i \nabla^2 \eta_i^n \Big) \quad (6)$$

for $i = 1, 2, ..., N$ grains with $\eta_i^{n+1} = \eta_i(t_{n+1})$, $\eta_i^n = \eta_i(t_n)$, and $\Delta t = t_{n+1} - t_n$.

In the equations describing the model, η is dimensionless, L_i has the dimensions $[L^2 T^{-1}]$, κ_i has the dimensions $[ML^2T^{-2}]$, F has the dimensions of energy $[ML^2T^{-2}]$, and f has the dimensions of energy density as $[ML^{-1}T^{-2}]$. All variables of the model are treated in a non-dimensional form in the simulation.

2 Prompt

The numerical implementation of the PF method based on the above equation is done using a Python code generated by GPT-4 on GPT Plus (GPT-4 in the following for convenience). Utilization of prompt engineering practices is done by breaking down the problem into sub-tasks with clear instructions and parameters. The overall prompt is broken into four major steps:

- System
- Context
- Specific instructions for numerical implementation
- Initialization cases
  - Case I : Ideal grain growth
  - Case II : Voronoi tessellation based initialization

2.1 System

The system prompt is mainly aimed for the chatbot to adopt a persona that has a contextual understanding of the broad field and maintains consistency in all responses. Further, the system prompt can also be used to obtain a specified style of programming. We also enforce the chatbot to provide an error-free code by providing such a system prompt.

In response to the prompt shown in Fig. 1, ChatGPT 4 mentions the applications of the PF method. It further elaborates by providing some mathematical aspects of the PF method such as its dependence on differential equations, commonly used numerical methods, initialization of random grain structures, use of Matplotlib and possible ways of evaluating results. It also gives a general example task and briefly explains the steps involved. This response indicates that the large language model acknowledges the persona and has an understanding of the PF method.

Prompt 1
Your role: You are an expert in computational materials science. You can understand complex mathematical model of physical phenomena in materials, which are usually described in terms of differential equations. Your particular research focus is the modeling of grain growth by the phase-field method. You understand the phase-field method in depth and know existing research in its implementation on grain growth modeling. You can judge the results of a numerical method based on its outputs, specifically plots. Further, you also know the best ways to initialize random grain structures in 2D. Your current role requires you to generate Python programs that are error-free (logically, semantically and numerically). You know the best practices of using the library Matplotlib to generate publication-ready plots.

Fig. 1 System prompt to provide ChatGPT 4 chatbot a persona that is an expert in the domain of our problem, understands the numerical methods and enforces error-free code generation

2.2 Context

The model description for our problem setting is provided in the context prompt. It includes specific equations as mentioned in the introduction. It is observed that providing equations of the PF model (1), (2) and (3) in the LaTeX format ensures the mathematical correctness in the code implementation.

ChatGPT 4 provides a clear understanding of the terms of the equation in the response to prompt in the Fig. 2. It mentions some applications of the Allen-Cahn equation first. Then it explains the individual terms, i.e. the order parameter η_i, mobility coefficient L_i, free energy F, gradient energy coefficients κ_i and the local free energy density f. It reproduces the equations provided and then provides the simulation approach briefly with suggestions for numerical methods and their possible limitations.

2.3 Specific Instructions for Numerical Implementation

Now, specific instructions are provided in a concise prompt shown in Fig. 3. To obtain a Python code that can be used for any type of initialized domain, we first consider a rectangular domain with circular grains at random locations with radii between a specified range. The motive is to use this generated phase field method code for other modified initializations such as Voronoi tessellation.

Prompt 2
We will conduct a grain growth evolution using PF modeling. Understand the following model first.
Model description: We have a grain growth model based on the PF method. The evolution of order parameters is given by the Allen-Cahn equation:

$$\frac{\partial \eta_i}{\partial t} = -L_i \frac{\delta F}{\delta \eta_i}$$

where, η_i is the order parameter for the i-th grain, $L_i = 5.0$ is the mobility coefficient and F is the free energy. The 'Fan and Chen' free energy function:

$$F = \int_V \left[f(\eta_1, \eta_2, \ldots \eta_N) + \sum_i^N \frac{\kappa_i}{2} |\nabla \eta_i|^2 \right] dv$$

where, $\kappa_i = 0.1$ are the gradient energy coefficients of grains and f is the local free energy density. The specific form of f is given as:

$$f(\eta_1, \eta_2, \ldots \eta_N) = \sum_i^N \left(-\frac{A}{2}\eta_i^2 + \frac{B}{4}\eta_i^4 \right) + \sum_i^N \sum_{i \neq j}^N \eta_i^2 \eta_j^2$$

, is used with $A = B = 1$.

Fig. 2 Context prompt providing the details of the PF model used in the problem with specific equations written in LaTeX format

Care has been taken to make the instructions clear and precise to restrict ChatGPT 4 in making any assumptions on its own. An emphasis on the desired dimension of the outputs has been made in the instructions. For e.g. the initialized order parameters should be stored in an array of dimensions $(Nx, Ny, ngrains)$, where (Nx, Ny) specifies the grid size and $ngrains$ is the number of grains. The discretized equation for time-stepping (4) is also provided in LaTeX format to ensure mathematical correctness in its code implementation (Fig. 3).

Prompt 3
You have understood the PF model for grain growth provided to you in the previous prompt. Now you can implement it numerically by following these instructions.
1) **Initialization**: Rectangular grid of size Nx = Ny = 64 with grid spacing dx = dy = 0.5. Initialize the domain with circular grains. Each grain is centered at a random point within the domain and has a radius between [14, 24]. Consider 3 grains. The order parameters for each grain are initialized by an array of shape (Nx, Ny, ngrains), where ngrains is the number of grains. Each ith 2D sub-array of shape (Nx, Ny) specifies the order parameter at all grid points for the ith grain. All the grid points falling within the ith grain will have order parameter of 1.0. All points outside this grain will have order parameter of 0.0.
2) **Numerical implementation**: Consider periodic boundary conditions. Apply Neumann boundary conditions to the right edge of the domain such that they hinder the grain growth. The time integration is carried out using Explicit Euler forward time stepping scheme. We thus obtain the following discretized expression.

$$\frac{\partial \eta_i}{\partial t} = \frac{\eta_i^{n+1} - \eta_i^n}{\Delta t} = -L\Big(- A\eta_i^n + B(\eta_i^n)^3 + 2\eta_i^n \sum_{i \neq j}^{N} (\eta_i^n)^2 - \kappa_i \nabla^2 \eta_i^n \Big)$$

for $i = 1, 2, ..., N$ grains. Approximate the Laplacian operator by a five-point stencil of the finite difference method. Compute the derivative of the energy accurately. Use a time step of dt = 0.005 and number of steps nsteps = 100. Ensure that order parameters do not exceed the bounds [0.0001, 0.999]. In each time step iteration, calculate the volume fraction of each grain. To calculate the volume fraction of a grain, sum all the order parameters over the grid and divide the sum by the area of the domain. Store the volume fraction of all grains over time. If the volume fraction of a grain is less than the threshold of 0.001, it is considered extinct and skipped from subsequent time iterations. Check this condition for each grain and track their status by a list of 0 and 1, where 1 indicates that the grain still exists. Encapsulate this numerical implementation in a function 'phase_field_evolve', which outputs a list 'eta_hist' of length equal to the nsteps. Each kth element of 'eta_hist' is an array of shape (Nx, Ny, ngrains), specifying the order parameters over all grid points for all ngrains at the kth time instant. The function 'phase_field_evolve' also outputs the stored volume fractions over time and the list of grain status.
3) **Computing the area fraction**: For the kth element of shape (Nx, Ny, ngrains) from 'eta_hist', iterate through each 2D sub-array of (Nx, Ny) cor-

Fig. 3 Prompt with specific instructions for numerical implementation provided as a one-thrust prompt

responding to the ith grain. In each 2D sub-array, check the eta values at all grid points with respect to a threshold of 0.5. Count the number of grid points with an eta value greater than this threshold. Divide this count by the area of the grid 'Nx*Ny'. This is the area fraction of the *i*th grain. Store this value of area fraction for each *i*th grain over all the time steps. Use this data later, to plot the area fractions of grains.
4) **Visualization**: We want to plot the grain growth of all grains in one plot corresponding to each time instant. Do this by the following steps. For the kth element of the shape (Nx, Ny, ngrains) from 'eta_hist', all the 2D sub-arrays of shape (Nx, Ny) are added. This is done for all time steps, and the resulting scaled 2D arrays are stored in a list. We use this list of 2D arrays to plot heatmaps with 'viridis' cmap for all time instants that are multiples of an integer 'nprint' which is taken as an input from the user. Create a function that makes an animation to visualize grain growth over time. Include colorbar in the animation.
5) **Verification**: Plot the area fraction computed in step 3 versus time plot for all the grains.

Fig. 3 (continued)

Prompt for pseudocode generation of grain growth simulation
Based on the instructions and steps provided in the previous prompt and using the equations from model description, provide an algorithm for the grain growth simulation and output it as a LaTeX code using the algorithm environment.

Fig. 4 Prompt to generate pseudocode for the grain growth simulation based on the specific instructions provided

ChatGPT 4 takes a relatively long time to process the response and indicates that it is analyzing its code. After completion it generates a plot of area fractions versus time steps as the output of verification. This shows that GPT 4 analyzes its generated code prior to responding and also adds the generated code output plot in its response. The generated Python code is shown in the code Listing 3.1.

The update of the order parameter, and thus the grain growth over time steps is conducted by the operations mentioned in the algorithm 2 generated by the prompt shown in Fig. 4.

Algorithm 2: Grain Growth Simulation using Phase-Field Method and Euler Forward Method

Data: Initial phase fields $\eta_i(x, y, t = 0)$, $i = 1, \ldots, N$;
Parameters: $A = 1$, $B = 1$, κ_i for $i = 1, \ldots, N$;
Grid size Nx, Ny; Grid spacing Δx, Δy;
Time step Δt; Total simulation time $T = nsteps \times \Delta t$;
Number of grains $ngrains = 3$.
Result: Phase field evolution $\eta_i(x, y, t)$ for $t \in [0, T]$, Volume fractions over time, Grain status.

Initialization: Set $t \leftarrow 0$;
Create a grid of size $Nx \times Ny$ with spacing Δx and Δy ;
Initialize $ngrains$ and for each grain i, set the order parameter $\eta_i(x, y) = 1.0$ inside the grain and 0.0 outside ;
while $t < T$ **do**
  for *each grid point* (x, y) **do**
    for *each grain* $i = 1, \ldots, N$ **do**
      Compute the Laplacian $\nabla^2 \eta_i$ using finite differences:

$$\nabla^2 \eta_i = \frac{1}{\Delta x^2} \Big(\eta_i(x + \Delta x, y) + \eta_i(x - \Delta x, y) + \eta_i(x, y + \Delta y) + \eta_i(x, y - \Delta y) - 4\eta_i(x, y) \Big)$$

      Update the order parameter using the Explicit Euler method:

$$\eta_i^{n+1} = \eta_i^n + \Delta t L \left(-A\eta_i^n + B(\eta_i^n)^3 + 2\eta_i^n \sum_{j \neq i}^{N} (\eta_j^n)^2 - \kappa_i \nabla^2 \eta_i^n \right)$$

      Enforce the bounds on η_i^{n+1} to keep it within [0.0001, 0.999] ;
    end
  end
  for *each grain* $i = 1, \ldots, N$ **do**
    Compute the volume fraction as the sum of η_i^{n+1} over all grid points divided by the area $Nx \times Ny$;
    if *volume fraction* < 0.001 **then**
      Mark grain i as extinct in the grain_status list ;
    end
  end
  Store η_i^{n+1} for all grains in eta_hist ;
  Apply boundary conditions;
  Update time: $t \leftarrow t + \Delta t$;
end
Post-processing: Analyze the final phase field distributions $\eta_i(x, y, t = T)$;
Output: η_{hist}, Volume fractions over time, Grain status.

2.4 *Initialization Cases*

The Python code obtained from the prompts so far is used along with two different initializations. The first case of ideal grain growth is considered for verification of the Python code for the PF method. The second case is based on Voronoi tessellations as initial domains and represents a general application in materials science.

2.4.1 Case I: Ideal Grain Growth

Ideal grain growth is a special case of normal grain growth, where the grain boundary motion is driven only by the local curvature of the grain boundary. Here, a spherical grain embedded in a large second grain is considered (Fig. 5).

2.4.2 Case II : Voronoi Tessellation

For a more generalized application of the generated PF method code, the domain is initialized by a Voronoi tessellation. Initially, a general random seed based Voronoi tessellation is generated and the order parameters are initialized. Next, it is modified for a Voronoi tessellation with a gradient in the cell size. Specific instructions are provided to initialize the order parameters to eliminate any discrepancies.

Prompt 4
Provide a Python function based on following instructions. Initialize the order parameters etas using following instructions. Rectangular grid of size Nx = Ny = 64 with grid spacing dx = dy = 0.5. We want to initialize a circular grain embedded in another large grain of the size of the rectangular grid. The circular grain of radius 14*dx* is centered at the center of the grid. Grain 1 is the larger grain and grain 2 is the embedded circular grain. The order parameters for each grain are initialized by an array of shape (Nx, Ny, ngrains), where ngrains = 2 is the number of grains. Each ith 2D sub-array of shape (Nx, Ny) specifies the order parameter at all grid points for the ith grain.
Initialization of grain 1: The order parameters over the rectangular grid for grain 1 are initialized to 1. But all points falling within the radius of the embedded circular grain have order parameter 0.
Initialization of grain 2: The order parameters over the rectangular grid for grain 0 are initialized to 0. But all points falling within the radius of the embedded circular grain have order parameter 1.

Fig. 5 Prompt for ideal grain growth initialization

Prompt 5
Create a Voronoi tessellation in a domain of size 32x32. Consider 25 grains. Ensure periodic boundary conditions. Extend any Voronoi cells near the domain boundary till the domain boundary such that we have the domain completely filled with the Voronoi cells. Plot the Voronoi tessellation.
After creating this Voronoi tessellation, initialize the order parameters for the PF method in the grains that lie in the trimmed Voronoi tessellation domain. Follow these steps.

1. Input: The Voronoi tessellation and its data from generated Voronoi tessellation. Create a rectangular grid of size 64 x 64 with grid spacing of 0.5.
2. Initialize Order Parameter: Create an array to store the order parameter values for each grain at each grid point. Create an array of dimensions (64, 64, ngrains), where 'ngrains' is the number of grains, i.e. the number of Voronoi cells.
3. For each grain, loop through all grid points. For each grid point (i, j) in the 64x64 grid, determine the corresponding Voronoi cell using the tessellation.
4. If the point is inside the Voronoi cell, set the order parameter for that point to 1.0. If the grid point is outside the Voronoi cell, set the order parameter for that point to 0.0.
5. Repeat for all grains in the Voronoi tessellation. Provide the final array of order parameters of dimensions (64, 64, ngrains).

Prompt 6
Modify the previous Voronoi tessellation and initialization of order parameters with a strongly decreasing Poisson point process. The gradient in the grain size should be easily noticeable. Ensure that the final array of etas has the same number of grains taken as an input.

Fig. 6 Prompts for Voronoi tessellation based initialization

3 Generated Code Listing and Results

The Python codes generated in response to the prompts mentioned in Sect. 2 are provided in this section with their output plots showing grain growth evolution.

3.1 Rectangular Domain with Randomly Located Circular Grains

The generated code after prompt 3 is displayed in the Code Listing 3.1. Therein, the number of steps was adapted manually to the target time period for grain evolution.

```
import numpy as np
import matplotlib.pyplot as plt
import matplotlib.animation as animation

# Parameters
Nx, Ny = 64, 64  # Grid size
dx, dy = 0.5, 0.5  # Grid spacing
ngrains = 3  # Number of grains
L = 5.0  # Mobility coefficient
A = B = 1  # Free energy parameters
kappa = 0.1  # Gradient energy coefficient
dt = 0.005  # Time step
nsteps = 1000  # Number of steps
threshold = 0.5  # Threshold for area fraction
radius_range = (7, 14)  # Range of radii for the grains
volume_fraction_threshold = 0.001  # Threshold for grain existence

# Initialize the domain with circular grains
def initialize_grains(Nx, Ny, ngrains, radius_range):
    eta = np.zeros((Nx, Ny, ngrains))
    centers = np.random.rand(ngrains, 2) * np.array([Nx, Ny])
    radii = np.random.randint(radius_range[0], radius_range[1], ngrains)

    for i in range(ngrains):
        for x in range(Nx):
            for y in range(Ny):
                if (x - centers[i, 0])**2 + (y - centers[i, 1])**2 <= radii[i]**2:
                    eta[x, y, i] = 1.0
    return eta

# Function to approximate Laplacian using a five-point stencil
def laplacian(eta, dx, dy):
    laplace_eta = (np.roll(eta, -1, axis=0) + np.roll(eta, 1, axis=0) - 2 * eta) / dx**2 + \
                  (np.roll(eta, -1, axis=1) + np.roll(eta, 1, axis=1) - 2 * eta) / dy**2
    return laplace_eta

# Phase field model evolution function
def phase_field_evolve(Nx, Ny, ngrains, dt, nsteps, eta_initial):
    eta_hist = [eta_initial]
    volume_fractions = np.zeros((nsteps, ngrains))
```

```
    grain_status = np.ones((nsteps, ngrains))

    for step in range(nsteps):
        eta_next = eta_hist[-1].copy()
        for i in range(ngrains):
            if grain_status[step-1, i] == 0:  # Skip extinct
    ↪  grains
                continue
            eta_i = eta_hist[-1][:, :, i]
            sum_eta_squared = sum([eta_hist[-1][:, :, j]**2
    ↪ for j in range(ngrains) if j != i])
            laplace_eta_i = laplacian(eta_i, dx, dy)

            # Discretized Allen-Cahn equation
            d_eta_dt = -L * (-A * eta_i + B * eta_i**3 + 2 *
    ↪  eta_i * sum_eta_squared - kappa * laplace_eta_i)
            eta_next[:, :, i] += d_eta_dt * dt

            # Enforce bounds on order parameters
            eta_next[:, :, i] = np.clip(eta_next[:, :, i],
    ↪ 0.0001, 0.999)

            # Calculate volume fraction
            volume_fraction = np.sum(eta_next[:, :, i]) / (
    ↪ Nx * Ny)
            volume_fractions[step, i] = volume_fraction
            if volume_fraction < volume_fraction_threshold:
                grain_status[step, i] = 0  # Mark grain as
    ↪ extinct


        eta_hist.append(eta_next)

    return eta_hist, volume_fractions, grain_status

# Function to compute area fraction for each grain at each
    ↪ time step
def compute_area_fraction(eta_hist, Nx, Ny, threshold):
    area_fractions = np.zeros((len(eta_hist), ngrains))

    for step, eta in enumerate(eta_hist):
        for i in range(ngrains):
            eta_i = eta[:, :, i]
            area_fraction = np.sum(eta_i > threshold) / (Nx
    ↪ * Ny)
            area_fractions[step, i] = area_fraction

    return area_fractions

# Initialize grains
eta_initial = initialize_grains(Nx, Ny, ngrains,
    ↪ radius_range)

# Run the phase field evolution
```

```
eta_hist, volume_fractions, grain_status =
    ↪ phase_field_evolve(Nx, Ny, ngrains, dt, nsteps,
    ↪ eta_initial)

# Compute area fractions
area_fractions = compute_area_fraction(eta_hist, Nx, Ny,
    ↪ threshold)

# Plotting area fraction vs. time for each grain
plt.figure(figsize=(10, 6))
for i in range(ngrains):
    plt.plot(np.arange(nsteps + 1), area_fractions[:, i],
    ↪ label=f'Grain {i+1}')
plt.xlabel('Time step')
plt.ylabel('Area fraction')
plt.title('Area fraction vs. Time for each grain')
plt.legend()
plt.show()

# create animation function was generated in a separate
    ↪ prompt response with minor modifications.
from matplotlib.colors import Normalize

import numpy as np
import matplotlib.pyplot as plt
import matplotlib.animation as animation
from matplotlib.colors import Normalize

def create_animation_with_colorbar(eta_hist, interval=100,
    ↪ nprint=10):
    """
    Create an animation visualizing the grain growth over
    ↪ time with a colorbar,
    using the squared sum of eta values and ensuring safe
    ↪ normalization.

    Parameters:
    - eta_hist: A list of numpy arrays, each representing
    ↪ the state of the system at a given time step.
    - interval: Time interval between frames in milliseconds
    ↪ .
    - nprint: Interval of time steps to update the animation
    ↪ .

    Returns:
    - anim: The Matplotlib animation object.
    """
    fig, ax = plt.subplots()
    ax.set_title('Grain Growth Over Time')

    # Compute the squared sum of eta values for the initial
    ↪ frame and normalize
    data = np.sum(eta_hist[0]**2, axis=2)
    max_val = np.max(data) if np.max(data) != 0 else 1  #
    ↪ Avoid division by zero
```

```
    data_normalized = data / max_val
    im = ax.imshow(data_normalized, animated=True, cmap='
↪ viridis', norm=Normalize(vmin=0, vmax=1))

    # Create colorbar
    fig.colorbar(im, ax=ax)

    def update(frame):
        """
        Update the plot for the animation, normalizing the
↪ data for each frame.
        """
        new_data = np.sum(eta_hist[frame]**2, axis=2)
        max_val = np.max(new_data) if np.max(new_data) != 0
↪ else 1  # Avoid division by zero
        new_data_normalized = new_data / max_val
        im.set_array(new_data_normalized)
        return (im,)

    anim = animation.FuncAnimation(fig, update, frames=range
↪ (0, len(eta_hist), nprint), interval=interval, blit=
↪ True)

    return anim

# This function is ready to be used as described in the
↪ comment at the bottom of the snippet.
# Usage example (assuming eta_hist is already computed):
anim = create_animation_with_colorbar(eta_hist, interval=50,
↪  nprint=10)
# To display in Jupyter Notebook:
from IPython.display import HTML
HTML(anim.to_html5_video())

# To save the animation as a file (uncomment and use in a
↪ local environment):
# anim.save('grain_growth_with_colorbar.mp4', writer='ffmpeg
↪ ', dpi=300)

# To save the animation as a file (uncomment and use in a
↪ local environment):
# anim.save('grain_growth_with_colorbar.mp4', writer='ffmpeg
↪ ', dpi=300)
```

Code Listing 3.1 Code generated by ChatGPT 4 after prompt 3. The code includes initialization of the domain and the numerical implementation as per steps specified in prompt 3

The output of the code creates a plot of area fractions with respect to time for all grains and an animation of the grain growth. The function that creates the animation was modified with minor changes to result in an animation with greater sharpness and to add a colorbar. The code also results a list of arrays having dimension $(Nx, Ny, ngrains)$ which stores the order parameters over the grid for all grains at all time instants. Moreover, the tracked grain status, area fractions and volume fractions are also obtained as output. Here, the initial $(t = 0)$, intermediate

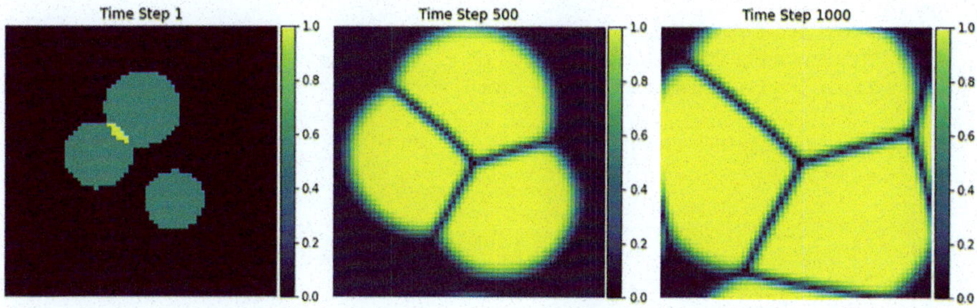

Fig. 7 Grain growth evolution for the generalized initialization in prompt 3 with three randomly located circular grains at initial, intermediate (step 500) and final state. The total simulation was conducted for 1000 time steps and a time-step size of 0.005. The domain size is 64×64 with a grid spacing of 0.5

($t = nsteps/2 \cdot dt$) and final state ($t = nsteps \cdot dt$) state of grain growth evolution is shown in Fig. 7. The final state here refers to the end of the time interval considered for the simulation.

3.2 *Ideal Grain Growth*

Case I with the ideal grain growth initialization was generated using the prompt 4. The resulting initialized order parameters were used in the PF method code listed in Code Listing 3.1.

```
import numpy as np
import matplotlib.pyplot as plt

def initialize_ideal_grain_growth(Nx, Ny, dx, dy,
    ↪ radius_factor, ngrains=2):
    """
    Initialize the order parameters for ideal grain growth
    ↪ with one circular grain
    embedded in a larger grain.
    Parameters:
    - Nx, Ny: Size of the rectangular grid.
    - dx, dy: Grid spacing.
    - radius_factor: Factor to determine the radius of the
    ↪ circular grain.
    - ngrains: Number of grains, default is 2.
    Returns:
    - eta: Initialized order parameters for each grain in
    ↪ the form of an array
           of shape (Nx, Ny, ngrains).
    """
    # Initialize the order parameter array
    eta = np.zeros((Nx, Ny, ngrains))
    # Calculate the center of the grid
```

```
    center_x, center_y = Nx / 2, Ny / 2
    # Calculate the radius of the circular grain
    radius = radius_factor * dx

    for x in range(Nx):
        for y in range(Ny):
            # Calculate the distance of each point from the center
            distance = np.sqrt((x - center_x)**2 + (y - center_y)**2)
            # Initialize grain 1 (the larger grain)
            if distance <= radius:
                eta[x, y, 0] = 0  # Inside the circular grain, set grain 1 to 0
                eta[x, y, 1] = 1  # Set grain 2 to 1
            else:
                eta[x, y, 0] = 1  # Outside the circular grain, set grain 1 to 1
    return eta

# Parameters for initialization
Nx, Ny = 64, 64  # Grid size
dx, dy = 0.5, 0.5  # Grid spacing
radius_factor = 28  # Multiplier for the radius based on dx

# Initialize the order parameters
eta_ideal_gg = initialize_ideal_grain_growth(Nx, Ny, dx, dy, radius_factor)

# Visualization
fig, axes = plt.subplots(1, 2, figsize=(12, 6))
titles = ['Grain 1 (Large Grain)', 'Grain 2 (Embedded Circular Grain)']
for i in range(2):
    im = axes[i].imshow(eta_ideal_gg[:, :, i], cmap='viridis', origin='lower')
    axes[i].set_title(titles[i])
    axes[i].axis('off')
    plt.colorbar(im, ax=axes[i])
plt.tight_layout()
plt.show()
```

Code Listing 3.2 Code generated by ChatGPT 4 after prompt 4 for initialization based on ideal grain growth case. The domain is initialized with a spherical grain embedded in a larger grain. The code results in an array of order parameters based on this initialization

The output of the generated PF code was an animation. Here, we show initial ($t = 0$), intermediate ($t = nsteps/2 \cdot \mathrm{d}t$) and final state ($t = nsteps \cdot \mathrm{d}t$) of the ideal grain growth case is shown in Fig. 8.

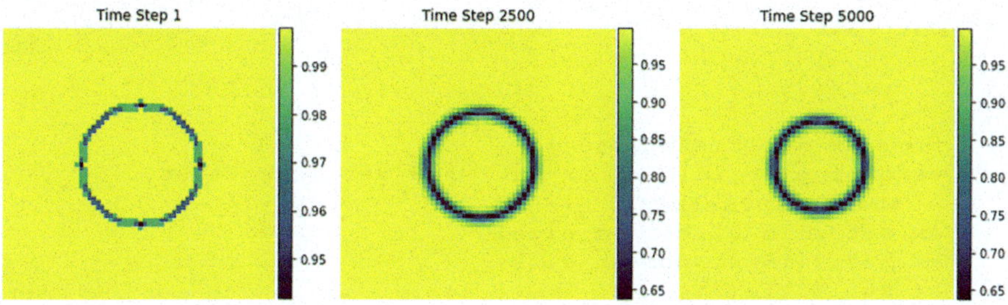

Fig. 8 Grain growth evolution for the ideal grain growth initialization as per prompt 4 with a spherical grain embedded in a larger cell. The evolution at initial, intermediate (step 2500) and final state shows shrinkage of the spherical cell. The total simulation was conducted for 5000 time steps and a time-step size of 0.005. The domain size is 64×64 with a grid spacing of 0.5

3.3 *Voronoi Tessellation Based Initialization*

Two types of Voronoi tessellations were considered for initialization. The first one has randomly distributed seeds following a uniform distribution. The second one has a decreasing gradient in the seeds, showing a variation in the grain size along the vertical axis. The corresponding generated codes are provided in code Listing 3.3 and code Listing 3.4.

```
import numpy as np
import matplotlib.pyplot as plt
from scipy.spatial import Voronoi, voronoi_plot_2d

def generate_voronoi(domain_size, ngrains):
    """
    Generate a Voronoi tessellation for a given domain size
    ↪ and number of grains.

    Parameters:
    - domain_size: The size of the domain.
    - ngrains: The number of grains to generate.

    Returns:
    - Voronoi tessellation object.
    """
    # Generate random points within the domain
    points = np.random.rand(ngrains, 2) * domain_size
    # Create the Voronoi tessellation
    vor = Voronoi(points)
    return vor

def plot_voronoi(vor, domain_size):
    """
    Plot a Voronoi tessellation, extending cells to the
    ↪ domain boundaries.

```

```
    Parameters:
    - vor: Voronoi tessellation object.
    - domain_size: The size of the domain.
    """
    fig, ax = plt.subplots()
    voronoi_plot_2d(vor, ax=ax, show_vertices=False,
        show_points=False)
    ax.set_xlim(0, domain_size)
    ax.set_ylim(0, domain_size)
    ax.set_title('Voronoi Tessellation with Extended Cells')
    plt.show()

def initialize_order_parameters(vor, grid_size, dx, ngrains)
    :
    """
    Initialize the order parameters for the phase field
    method based on Voronoi tessellation.

    Parameters:
    - vor: Voronoi tessellation object.
    - grid_size: The size of the grid.
    - dx: The grid spacing.
    - ngrains: The number of grains.

    Returns:
    - Initialized order parameters array.
    """
    eta = np.zeros((grid_size, grid_size, ngrains))
    for i in range(grid_size):
        for j in range(grid_size):
            # Find the closest Voronoi cell for each grid
    point
            x, y = (i + 0.5) * dx, (j + 0.5) * dx
            distances = np.sqrt((vor.points[:, 0] - x)**2 +
    (vor.points[:, 1] - y)**2)
            closest = np.argmin(distances)
            eta[i, j, closest] = 1.0
    return eta

# Parameters
domain_size = 32
ngrains = 25
grid_size = 64
dx = 0.5

# Generate and plot Voronoi tessellation
vor = generate_voronoi(domain_size, ngrains)
plot_voronoi(vor, domain_size)

# Initialize order parameters based on the Voronoi
    tessellation
eta = initialize_order_parameters(vor, grid_size, dx,
    ngrains)
```

```
# eta contains the order parameters array of dimensions (64, 64, 25)
```

Code Listing 3.3 Code generated by ChatGPT 4 after prompt 5 for the general Voronoi tessellation based initialization. The code results in an array of order parameters based on this initialization and a plot of the Voronoi tessellation

```
import numpy as np
import matplotlib.pyplot as plt
from scipy.spatial import Voronoi, voronoi_plot_2d

def generate_strongly_decreasing_points(domain_size, ngrains):
    """
    Generate points for a Poisson Voronoi tessellation with a strongly decreasing density,
    ensuring a noticeable gradient in grain size across the domain.
    """
    points = []
    while len(points) < ngrains:
        x = np.random.uniform(0, domain_size)
        y = np.random.uniform(0, domain_size)
        # Use a non-linear decreasing function for point density
        probability = (domain_size - x)**2 / domain_size**2  # Quadratically decreasing probability
        if np.random.rand() < probability:
            points.append([x, y])
        if len(points) > ngrains:  # Reset if over-generated
            points = []

    return np.array(points[:ngrains])

def initialize_order_parameters_exactly(grid_size, dx, points, ngrains):
    """
    Initialize the order parameters for the phase field method, ensuring the specified number of grains.
    """
    eta = np.zeros((grid_size, grid_size, ngrains))

    for i in range(grid_size):
        for j in range(grid_size):
            pos = np.array([i * dx, j * dy])
            distances = np.linalg.norm(points - pos, axis=1)
            closest_point_index = np.argmin(distances)
            eta[i, j, closest_point_index] = 1.0

    # Ensure eta array reflects exactly 'ngrains' grains
    assert eta.shape[2] == ngrains, "The number of grains in eta does not match ngrains."

    return eta

# Parameters remain the same as the previous example
```

```
domain_size = 32
ngrains = 25
grid_size = 64
dx = dy = 0.5

# Generate points with a strongly decreasing process
points_strong_gradient = generate_strongly_decreasing_points
    ↪ (domain_size, ngrains)

# Create the Voronoi tessellation
voronoi_strong_gradient = Voronoi(points_strong_gradient)

# Initialize the order parameters
eta_strong_gradient = initialize_order_parameters_exactly(
    ↪ grid_size, dx, points_strong_gradient, ngrains)

# Plot the Voronoi tessellation with a strong gradient
fig, ax = plt.subplots()
voronoi_plot_2d(voronoi_strong_gradient, ax=ax,
    ↪ show_vertices=False)
ax.set_xlim(0, domain_size)
ax.set_ylim(0, domain_size)
ax.set_title("Poisson Voronoi Tessellation with Strong
    ↪ Gradient")
plt.show()
```

Code Listing 3.4 Code generated by ChatGPT 4 after prompt 6 for the initialization with Voronoi tessellation having a decreasing gradient. The code results in an array of order parameters based on this initialization and a plot of the Voronoi tessellation. ChatGPT 4 assumes a suitable non-linear function for the point density on which the seeds of the Voronoi tessellation are created

Initial ($t = 0$), intermediate ($t = nsteps/2 \cdot dt$) and final state ($t = nsteps \cdot dt$) of the Voronoi tessellation based grain growth in both cases is shown in Figs. 9 and 12 respectively.

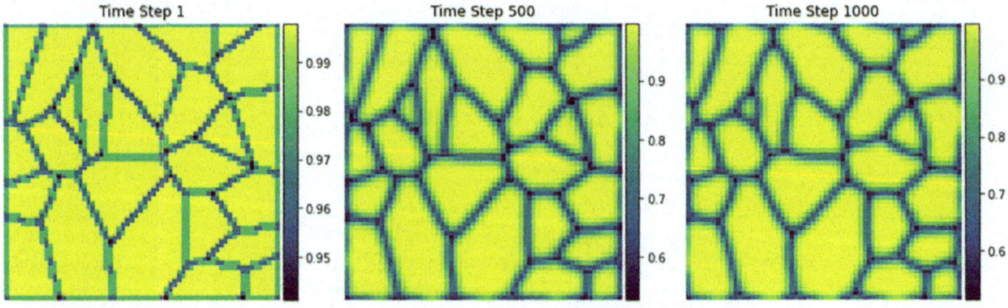

Fig. 9 Grain growth evolution for the general Voronoi tessellation based initialization as described in prompt 5. The evolution at initial, intermediate (step 500) and final state is shown in the plots. The total simulation was conducted for 1000 time steps and a time-step size of 0.005. The domain size is 64×64 with a grid spacing of 0.5

4 Tests for Verification

In the current problem setting, errors that do not allow the code to run, root from index errors and array dimension errors. However, using ChatGPT 4 and adhering the prompt engineering practices, it is observed that the generated code from the prompts runs without such errors. However, the logical and mathematical correctness of numerical implementation needs to be thoroughly verified. We discuss these verification aspects in three stages:

- Initialization
- Numerical implementation
- Verification with expected physical observations

Minor errors in the specifics of plotting and visualization are easily rectified by instructing the chatbot in subsequent prompts. For example, the function that creates the animation `create_animation_with_colorbar` was modified with minor changes to result in an animation with greater sharpness and add to a colorbar.

4.1 Initialization

The two initialization cases considered are first checked. First, the dimensions of the output array of initialized order parameters are checked. These should be $(Nx, Ny, ngrains)$, where (Nx, Ny) signifies the grid size of the domain and $ngrains$ is the number of grains, both provided as an input. Next, individual heatmaps of some of the initialized grains are plotted and verified visually. Figure 10 shows the initialization in case of ideal grain growth as per the prompt in Sect. 2.4.1.

4.2 Numerical Implementation

Even though the codes run without errors, they can still have logical and mathematical errors in terms of the numerical implementation. These are checked by the following steps:

- Verifying if the initialized arrays to store updated variables are of correct dimensions.
- Checking the correct implementation of the five-point stencil to approximate the Laplacian operator. In some trials, it is observed that division with the square of step size was missing in the generated code.
- Checking the mathematically correct implementation of the discretized time-stepping equation (4). Specifically, the signs in front of each individual term need to be checked, and ensuring that the sum in the interaction term, i.e. $\sum_{i \neq j}^{N} (\eta_i^n)^2$, has

been correctly evaluated. There were instances where the condition $i \neq j$ in the summation operator was not strictly followed, and instead overall sum of all η_j^2 was resulted. Such issues were overcome by providing the discretized time-stepping equation in a LaTeX format in the prompt.

- Checking if volume fraction is correctly computed and the condition for extinction of grains is followed.
- Checking if the bounds on order parameters are enforced.
- Checking if any specified boundary conditions are imposed. Surprisingly, ChatGPT 4 omitted the line in prompt 3, instruction 2 (refer Fig. 4) which instructs imposing Neumann boundary conditions on the right edge of the domain. However, this can be easily rectified by instructing ChatGPT 4 in a subsequent prompt.
- Checking computation of area fractions.
- Verifying that the parameters mentioned in the prompt are correctly assigned before running the specific functions of the code.

4.3 Verification with Expected Physical Observation

Ideal grain growth using the code generated by ChatGPT 4 can be used for verification against physical observations. For the shrinking spherical grain embedded in a large second grain, it is known that the change in radius of the shrinking grain can be approximated as:

$$D^2 - D_0^2 = kt \tag{7}$$

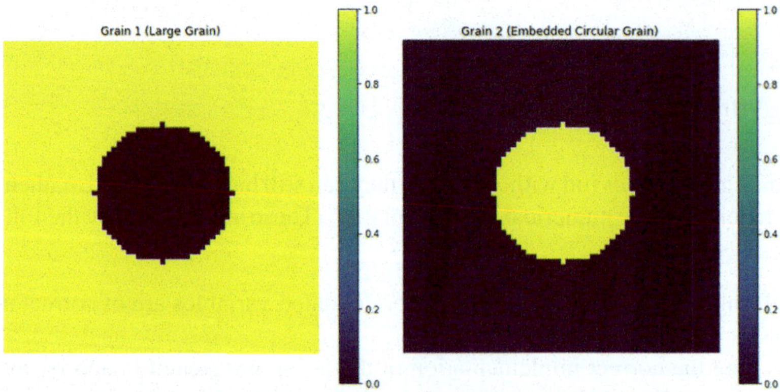

Fig. 10 Verifying initialization for ideal grain growth case by plotting heat-maps of individual grains from the initialized order parameter array. The heat-maps show the desired initialization of the spherical grain 2 embedded in the larger grain 1

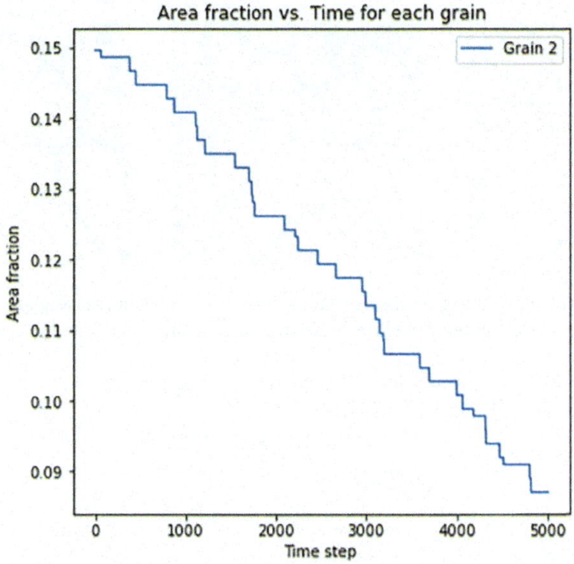

Fig. 11 Area fraction versus time step plot for ideal grain growth case

where, D and D_0 are the current and the initial grain radius, respectively, and k is a temperature-dependent constant. A similar behavior can be observed by the plot of area fractions versus time steps using the computation by the generated code. This is shown in the Fig. 11, where the spherical grain has a radius of $28\mathrm{d}x$, where $\mathrm{d}x$ is the grid spacing and the evolution is carried out with a time step size of 0.005 for 5000 steps.

Similarly, for the Voronoi tessellation based initialization, it is observed that the evolution follows the physical observation that larger grains grow and smaller ones disappear eventually. This can be observed in the Fig. 12. The same is also reflected in the area fraction versus time step plot as shown in the Fig. 13. The area fraction of the two small grains with indices 3 and 6 decreases to zero as shown in the plot lines with color red and pink respectively.

5 Discussion

In the context of code generation, ChatGPT 4 performs surprisingly well with a minimal amount of errors. However, as the complexity of tasks increases, the chances of running into errors, specifically, logical errors, increase. In such cases, the large language model (LLM) produces *hallucinations*, which are outputs that do not match the intent of the prompt. To drive the LLM to a desired output for a complex task, a suitably tailored prompt has to be created. In this regard, a suitable prompt design

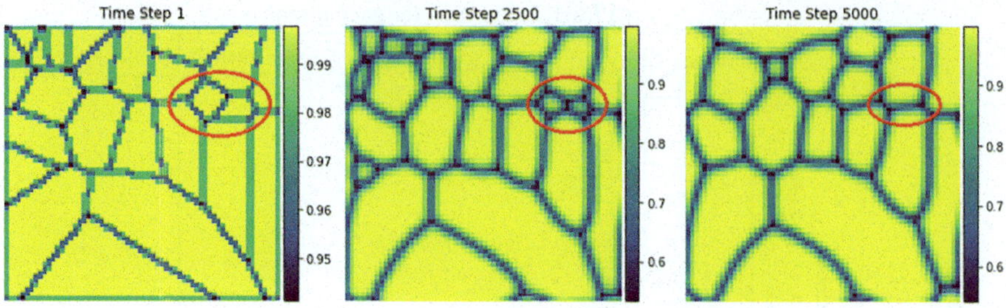

Fig. 12 Grain growth evolution for the Voronoi tessellation with a decreasing gradient as described in prompt 6. As indicated in the red ellipse, the two small grains shrink and eventually vanish

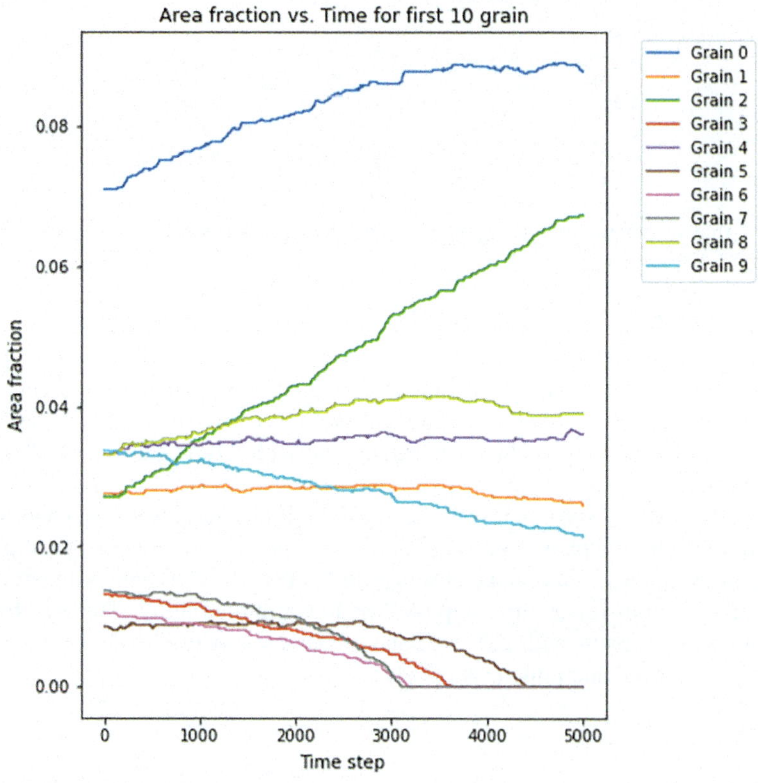

Fig. 13 Area fraction versus time step plot for the first 10 grains in the Voronoi tessellation with a decreasing gradient as described in prompt 6. The area fractions of the two small grains with indices 3 (red) and 6 (pink) decrease and reach zero indicating that these grains have disappeared. The larger grains growth with varying growth rates

is important. Here, we discuss the key implications of prompt engineering tactics in the context of the current problem setting. The following prompt engineering tactics are in the prompt design:

- **Write clear instructions**: Clarity in instructions provides limited room for the LLM to make its own assumptions and in worse cases, produce hallucinations. However, including too many unnecessary details is not suitable. This was noticed when details of the Voronoi tessellation algorithm and the specific programming steps were mentioned in a detailed prompt to create Voronoi tessellation based initialization. Such details caused programming errors such as those related to indexing and broadcasting. When only limited instructions were given with certain restrictions such as periodic boundary conditions and a completely filled domain, as shown in prompt 5 in Fig. 6, the LLM resulted in an error free code with the desired outputs. This shows that the prompts should include specific instructions that build upon the logic of solving the problem. But the specifics of programming the instructions into a code need not be included.
- **Split complex tasks into simpler subtasks**: Breaking complex tasks into sub-steps is a good approach to provide the LLM with the rigorous logic of solving the problem. Instead of sequential prompting, one can also concatenate all the steps concisely as done in prompt 3 shown in Fig. 3. Existing functions generated in response to previous prompts in one chat can be modified for different cases, such as done in the case of prompt 6 which modifies the function generated from prompt 5 shown in Fig. 6. However, recalling code or information from prompt responses at an initial stage in a long chat can cause errors due to the limited short-term memory of the model. This can be remedied by restating the code or information from such a previous response in the new prompt. This also helps in maintaining the consistency of the code variables and functions, as we proceed with sequential subtasks.
- **Ask model to adopt a persona**: It is observed that providing a system prompt as shown in Fig. 1, improves the generation of relevant responses and maintains consistency in the style of code generation. Moreover, specific instructions to obtain a certain style of response can also be added to a system prompt if needed.
- **Providing necessary equations**: It is important to provide the context of the problem in consideration through the required equations. This enables the LLM to develop an understanding of the problem to be solved and improve the relevance of the responses to the problem. In our case, we provide all the equations of the model description in prompt 2 shown in Fig. 2, since the free energy function as defined in (2) is not a standard well-known differential equation like Laplace equation in heat transfer or Burgers' equation in fluid dynamics. In our case, without specifying the equations in model description, ChatGPT 4 assumed some equations which did not exactly resemble to the equations describing the PF model.

  ChatGPT 4 could also perform the relatively complex mathematical operation of differentiating the free energy function (2) and local free energy density (3) and use it in the non-conserved Allen-Cahn equation (1). Further, it implemented the specified numerical methods to obtain the discretized Eq. (4) for the time-

stepping of order parameters. Despite correct mathematical operations, sometimes it made errors in maintaining the mathematical correctness during code generation as mentioned in the verification of numerical implementation in Sect. 4.2. It is also important to note that providing equations in the LaTeX format ensured their correct interpretation and mathematical understanding by the LLM.
- **To do or not to do?**: It is observed that prompts with statements that tell "what not to do" result in more erroneous responses as compared to prompts that avoid them. This has been also followed in the prompts used in the current problem. For example, in prompt 5 shown in Fig. 6, instead of mentioning, "Do not let the domain be empty at any location.", it is mentioned in a specific way with a remedy to the issue by stating, "Extend any Voronoi cells near the domain boundary till the domain boundary such that we have the domain completely filled with Voronoi cells."

Designing a prompt that adheres to these tactics is an iterative process and may require some trials. It is usually not prudent to continue with one chat for multiple variations of the same prompt. Furthermore, rectifying errors generated by a sub-optimal prompt can help in cases of few errors. However, if the generated code contains several syntax, semantic or logical errors, then it is suggested to redraft a better prompt and use it in a new chat. This iterative process of prompt design also signifies the importance of human intervention in using ChatGPT 4 for code generation.

6 Conclusion

ChatGPT 4 has been used to generate a PF code for grain growth evolution. The Fan and Chen [6] model has been applied and the equations describing the model have been solved numerically using the explicit Euler forward method along with the five-point stencil of the finite difference method for approximating the Laplace operator. The complex problem has been broken into steps that instruct ChatGPT 4 to generate the code. To this end, specific prompt engineering tactics have been used to design prompts that generate a PF method code which can be applied to generalized initialization of the order parameters. Specific applications of this code are shown by developing two initialization cases, (1) ideal grain growth with a spherical grain embedded in a larger grain and (2) Voronoi tessellations. The initialization has also been carried out by code generated by ChatGPT 4 using suitable prompts. The resulting code generated by ChatGPT 4 is error-free in terms of syntax, semantic and logical errors. Mathematical correctness of the numerical implementation has been ensured by rigorous checking of the generated code. The final results have been verified with expected results based on general physical observations in grain growth evolution. Details of the prompt engineering tactics applied in the design of the prompt have been explained.

In extension to this work, newer prompt engineering methods like zero-shot chain of thought prompting [8] can be used, which allow the LLM to generate its own reasoning for solving the problem. This reasoning can be used as an initial point for suitable prompt design. Another approach to solving complex code generation tasks is to provide the steps to solve the task in terms of a code template with comments. The aim is to give the LLM some direction to solve the problem as well as a template to structure the code, and allow it to develop detailed solutions with code based on this limited direction. In case of complex initialization, such as a specific grain microstructure, its image can be provided as an input in the prompt, with required conditions, and ask ChatGPT 4 to generate a code that provides a similar initialized microstructure. With the presence of LLMs like ChatGPT 4 and their continuous developments, prompt engineering based code generation will become an indispensable approach for computational materials scientists to tackle complex problems.

References

1. P. Blikstein and A. P. Tschiptschin. Monte Carlo simulation of grain growth. *Materials Research*, 2:133–137, 1999.
2. F. Wakai, N. Enomoto, and H. Ogawa. Three-dimensional microstructural evolution in ideal grain growth-general statistics. *Acta Materialia*, 48(6):1297–1311, 2000.
3. H. J. Frost and C. V. Thompson. Computer simulation of grain growth. *Current Opinion in Solid State and Materials Science*, 1(3):361–368, 1996.
4. Y. Liu, T. Baudin, and R. Penelle. Simulation of normal grain growth by cellular automata. *Scripta materialia*, 34(11), 1996.
5. S. Biner. Solving Phase-Field Models with Finite Difference Algorithms. *Programming Phase-Field Modeling*. pp. 17-97 (2017),
6. D. Fan and L.-Q. Chen. Computer simulation of grain growth using a continuum field model. *Acta Materialia*, 45(2):611–622, 1997.
7. D. Weygand, Y. Brechet, and J. Lepinoux. A vertex dynamics simulation of grain growth in two dimensions. *Philosophical Magazine B*, 78(4):329–352, 1998.
8. T. Kojima, S. S. Gu, M. Reid, Y. Matsuo, and Y. Iwasawa. Large language models are zero-shot reasoners, 2023.

Modeling Corrosion Using a Cellular Automaton

Mehdi Bakhshi Zadeh and Bernhard Eidel

Abstract This study develops a 2D cellular automaton model to simulate corrosion processes using Fick's second law of diffusion and Reaction Kinetics. The model accurately predicts corrosion initiation and propagation under various environmental conditions, providing valuable insights for designing effective corrosion prevention strategies. The verification process confirmed the accurate implementation of theoretical principles, while simulation results highlighted the progression of corrosion and the impact of protective layers and different metal corrosion rates. Furthermore, the model emphasizes the potential of AI-assisted tools in advancing research methodologies. Continuous refinement and expert oversight are essential to enhance the reliability of these tools. This work demonstrates the significant role of Cellular Automata (CA) in understanding complex corrosion mechanisms and underscores the necessity for ongoing improvements in simulation techniques. By integrating theoretical and practical aspects, this study offers a robust framework for future research and applications in corrosion management.

1 Introduction

Corrosion is a natural, inevitable process that results in the gradual deterioration of materials, particularly metals, through chemical or electrochemical reactions with their environment. This phenomenon is driven by the tendency of materials to return to their more stable, lower-energy states, often leading to the formation of oxides or other compounds. The consequences of corrosion are significant, affecting the structural integrity, safety, and longevity of infrastructure, machinery, and industrial

M. Bakhshi Zadeh (✉) · B. Eidel
M5, IMFD, TUBAF, Freiberg, Germany
e-mail: Mehdi.Bakhshi-Zadeh@student.tu-freiberg.de

B. Eidel
e-mail: bernhard.eidel@imfd.tu-freiberg.de

© The Author(s), under exclusive license to Springer Nature Switzerland AG 2025
B. Eidel (ed.), *GPT for Python-Coding in Computational Materials Science and Mechanics*, Studies in Computational Intelligence 1198,
https://doi.org/10.1007/978-3-031-85470-5_4

components. In the absence of effective management strategies, corrosion can lead to catastrophic failures, causing economic losses and safety hazards.

Corrosion can occur uniformly across a material's surface or in localized forms, such as pitting and crevice corrosion, which lead to concentrated damage in specific areas. Pitting corrosion, in particular, is a dangerous form as it can lead to the rapid penetration of materials with minimal overall mass loss. Understanding the mechanisms behind various corrosion types is vital for developing effective prevention and mitigation strategies, which can significantly reduce the risk of structural failures.

Recent advancements in computational modeling, such as Cellular Automata (CA), have provided powerful tools to simulate and understand the complex, dynamic nature of corrosion processes. CA provides a discrete modeling framework, enabling the simulation of corrosion at various scales, from the initiation of corrosion pits to the evolution of surface roughness in materials. This study focuses on applying CA models, particularly through Fick's second law and reaction kinetics, to simulate and predict corrosion processes in a 2D environment.

Studies show that CA models are effective in capturing the stochastic nature of localized corrosion, such as pitting, in various materials including low-carbon steels [4]. Additionally, CA has been employed to model oxidation mechanisms in steel, accurately simulating the diffusion of oxygen and the formation of oxide layers in high-temperature environments [5]. These applications highlight the versatility of CA as a tool for studying a wide range of corrosion scenarios across different materials and environmental conditions.

1.1 Theoretical Backbone

Cellular Automata (CA) are computational models that utilize discrete grid systems to simulate the evolution of complex systems over time. Each cell within the grid can exist in a finite number of states and interact with its neighboring cells based on predefined rules. The classification of cells in the corrosion process is illustrated in Fig. 1. This figure shows how each cell transitions during the corrosion process, highlighting the different states and interactions that are critical to the simulation model. The diagram helps to visualize the cellular automaton framework and underscores

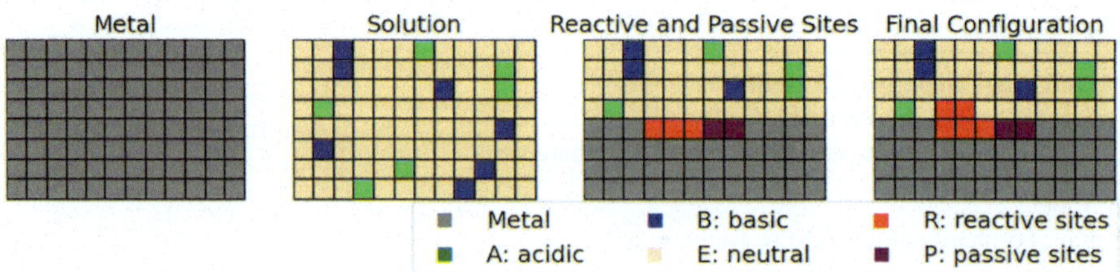

Fig. 1 The cell classification model and how each cell changes in the corrosion process

the importance of accurate state classification in predicting the progression of corrosion. Key ingredients of CA include the grid structure, state set, neighborhood, and transition rules. This method is particularly effective in modeling local interactions and emergent global behaviors, making it suitable for simulating corrosion processes where localized interactions can lead to significant material degradation. The utility of this approach in corrosion management has been well documented. For instance, Wang et al. (2019) demonstrated the effectiveness of CA models in simulating the high-temperature corrosion of Ni-based alloys in chloride molten salts, showing how these models can predict the growth of corrosion layers and the migration of elements under various conditions [1].

From an electro-chemical perspective, corrosion involves anodic reactions where metal atoms lose electrons and form metal ions (e.g., $\mathrm{Fe} \rightarrow \mathrm{Fe}^{2+} + 2e^-$), and cathodic reactions where electrons are consumed, typically by a reduction process (e.g., $\mathrm{O}_2 + 4H^+ + 4e^- \rightarrow 2H_2O$). The presence of an electrolyte is crucial as it facilitates ion and electron movement, and the electrochemical cells created by these reactions drive the corrosion process through the potential difference they generate. The approach further highlights the ability of CA models to capture intricate interactions between electrochemical reactions, material properties, and environmental factors, providing valuable insights into corrosion mechanisms and aiding in the development of predictive maintenance strategies [2]. Additionally, CA models allow for the incorporation of stochastic elements, which are essential in capturing the inherent randomness and complexity of corrosion processes. This makes CA an invaluable tool for researchers and engineers who aim to develop predictive models and design effective corrosion prevention measures.

Further studies have emphasized the application of CA models in simulating different types of corrosion phenomena. For example, Chen and Wen [7] utilized CA models to simulate the uniform corrosion damage evolution of steel structures exposed to acid rain, demonstrating how varying concentrations of corrosive agents impact the corrosion depth over time. Similarly, Xiao et al. [8] applied CA models to simulate pitting corrosion in Ni-based alloys, showing the influence of chloride ion concentration and solution pH on the evolution of pits. These studies underscore the versatility of CA in capturing both uniform and localized corrosion processes, offering detailed insights that are crucial for predicting material degradation in various environments.

The model implemented in this study employs two critical theoretical components: Fick's Second Law (FSL) of Diffusion and reaction kinetics. FSL describes how the concentration of a substance changes over time due to diffusion, which is fundamental in modeling the transport of corrosive species in materials. The equation is given by:

$$\frac{\partial C}{\partial t} = D\nabla^2 C \,, \tag{1}$$

where C is the concentration, t is time, and D is the diffusion coefficient. This equation is discretized and applied within the CA framework to simulate the diffusion of corrosive agents. The implementation of FSL in CA involves discretizing the spatial domain into a grid where each cell represents a concentration of the diffusing species and updating the concentration based on the diffusion coefficient and the concentration gradient. This approach allows for a detailed representation of the spatial and temporal evolution of the concentration field, capturing the effects of diffusion on the corrosion process.

Reaction kinetics (RK) are incorporated to model the chemical reactions occurring at the material's surface, which contribute to the corrosion process. The rate of these reactions is governed by factors such as temperature, concentration of reactants, and the presence of catalytic agents. In our CA model, these reactions are represented through state transitions of the cells, where the probability of transition depends on the local concentration of reactive species. For example, the interaction between the diffusing oxygen and metal atoms to form oxides can be modeled as a probabilistic state transition [3]. This probabilistic approach ensures that the model can capture the inherent variability and stochastic nature of corrosion reactions.

In the context of high-temperature corrosion, the diffusion of oxygen and chloride ions through molten salt layers and their reaction with metal surfaces is critical. Wang et al. [1] provides a detailed framework for modeling these interactions, showing that the formation of protective layers and the degradation of the metal substrate can be effectively simulated using CA. This detailed framework is essential for understanding how different environmental conditions and material properties influence the corrosion process, enabling the development of targeted strategies for corrosion mitigation.

1.2 Problem Statement

This study aims to address the challenge of predicting and managing corrosion in industrial applications by utilizing a 2D CA model. The specific problem involves simulating the diffusion of corrosive species and the resultant chemical reactions using FSL and reaction kinetics. The parameters used in the corrosion simulation are summarized in Table 1.

In the following, we provide the prompt used to generate the algorithm for the corrosion modeling by ChatGPT-4. This prompt is specifically used to instruct the AI to create an algorithm based on FSL of diffusion and reaction kinetics within a 2D cellular automaton framework.

The algorithm as LaTeX-pseudocode, which is generated by the prompt in Fig. 2 is given in the algorithm box 3.

Table 1 Parameters used in the corrosion simulation study

Parameters	Values/Types	Dimensions
Domain size (N_x, N_y)	200	$[L]$
Protection layer thickness	2	$[L]$
Initial damage (crack) indices	List of tuples	$[L] \times [L]$
Diffusion coefficient (top, D_{top})	0.1	$[L^2T^{-1}]$
Diffusion coefficient (bottom, D_{bottom})	0.05	$[L^2T^{-1}]$
Time step size (Δt)	0.01	$[T]$
Spatial step size (Δx, Δy)	1	$[L]$
Random factor (random_factor)	Uniform distribution	[1]
Output	Corrosion animation	–

Prompt:
You are an expert in corrosion simulation based on Fick's second law of diffusion and reaction kinetics, including both anodic and cathodic parts. For this simulation, a cellular automaton is used to model the corrosion process in 2D. Please generate the LaTeX code for this algorithm using the `algorithm2e` package.
The algorithm should:

- Accept input parameters including grid size, time steps, time step size, grid spacing, diffusion coefficient, initial concentration, reaction rates, and threshold potential.
- Initialize the grid and concentration fields, as well as the corrosion state.
- Perform a diffusion step using Fick's second law, calculating the concentration at each grid point using the central finite difference scheme.
- Update the concentration based on corrosion reaction kinetics for both anodic and cathodic reactions.
- Use a cellular automaton to update the corrosion state, activating corrosion based on local concentration and neighboring states.
- Apply boundary conditions to the concentration and corrosion state fields.
- Finally, output the concentration field and corrosion state over time.

Please format this as a structured algorithm using the `algorithm2e` package in LaTeX.

Fig. 2 Prompt for generating LaTeXpseudocode of an algorithm for corrosion simulation along with CA

Algorithm 3: Corrosion Modeling in 2D using Fick's 2nd Law and Cellular Automata

Input: Grid size (Nx, Ny), time steps T, time step Δt, grid spacing Δx, Δy, diffusion coefficient D, initial concentration $C(\boldsymbol{x}, 0)$, reaction rates k_a, k_c, threshold potential V_{th}

Output: Concentration field $C(\boldsymbol{x}, t)$ and corrosion state over time

Initialization:

Set the grid with size (Nx, Ny);

Initialize the concentration field $C(\boldsymbol{x}, 0)$;

Initialize the corrosion state $S(\boldsymbol{x}, 0)$ (1 for active corrosion sites, 0 otherwise);

Set diffusion coefficient D and reaction rates k_a, k_c;

for $t = 0$ **to** T *with step* Δt **do**

  Diffusion Step (using Fick's 2nd Law):

  foreach *grid point* (i, j) **do**

    Calculate the concentration $C(i, j)^{n+1}$ at the next time step using the central finite difference scheme:

$$C(i, j)^{n+1} = C(i, j)^n + D\Delta t \left(\frac{C(i+1, j)^n - 2C(i, j)^n + C(i-1, j)^n}{\Delta x^2} + \frac{C(i, j+1)^n - 2C(i, j)^n + C(i, j-1)^n}{\Delta y^2} \right)$$

  end

  Corrosion Reaction Kinetics:

  foreach *grid point* (i, j) **do**

    Calculate the local potential $V(i, j)^n$ based on the concentration $C(i, j)^n$ and other factors (e.g., electrolyte, material properties);

    if $V(i, j)^n > V_{th}$ ***and*** $S(i, j)^n = 1$ **then**

      Update concentration due to anodic reaction:

$$C(i, j)^{n+1} = C(i, j)^{n+1} - k_a \Delta t$$

    end

    else if $V(i, j)^n < V_{th}$ ***and*** $S(i, j)^n = 1$ **then**

      Update concentration due to cathodic reaction:

$$C(i, j)^{n+1} = C(i, j)^{n+1} + k_c \Delta t$$

    end

  end

  Cellular Automaton Update:

  foreach *grid point* (i, j) **do**

    Update the corrosion state $S(i, j)^{n+1}$ based on local concentration $C(i, j)^{n+1}$ and neighboring states:

    if $C(i, j)^{n+1} > C_{crit}$ ***or*** *neighboring sites* $S(i+1, j)$, $S(i-1, j)$, $S(i, j+1)$, $S(i, j-1)$ *are corroded* **then**

      $S(i, j)^{n+1} = 1$;

      `// Activate corrosion if concentration exceeds threshold or neighboring sites are corroded`

    end

  end

  Apply Boundary Conditions:

  Apply appropriate boundary conditions (e.g., Dirichlet, Neumann) to the concentration and corrosion state fields;

end

Output:

Return the concentration field $C(\boldsymbol{x}, t)$ and corrosion state $S(\boldsymbol{x}, t)$ over time;

2 Prompt

In the following, the sketch of the simulation domain in Fig. 3 with its features is introduced and explained as follows. The coordinate system, labeled with $\boldsymbol{x}$ and $\boldsymbol{y}$ axes, intersects at the origin (0, 0), defining the framework of the simulation. The domain consists of a 2D rectangular grid with dimensions $N_x \times N_y$ and grid spacing $\Delta \boldsymbol{x}$ and $\Delta \boldsymbol{y}$. Initial damage sites, highlighted in red at the top and bottom, indicate entry points for corrosive substances. Surrounding the grid, dashed blue lines represent a protective layer of specified thickness. The upper half of the grid, referred to as the Top Metal, is shaded light grey, while the lower half, the Bottom Metal, is shaded dark grey. Green arrows illustrate the diffusion of corrosive species through the grid. Orange patches and arrows depict the state transitions of cells from uncorroded to corroded states. Additionally, the "Time-stepping Loop" annotation illustrates the iterative update process of the simulation.

The initial prompt for generating the code is given in Fig. 4.

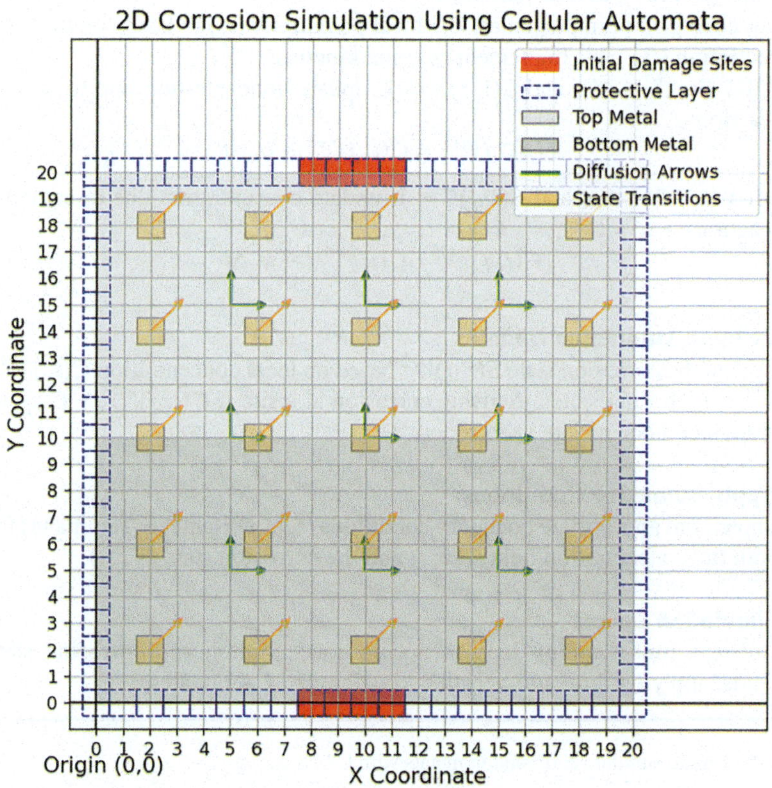

Fig. 3 Detailed sketch illustrating the 2D corrosion simulation process on metal surfaces using a Cellular Automaton, highlighting critical parts, diffusion paths, and state transitions

Define the Domain:
Create a 2D grid representing the metal surface with size $Nx \times Ny$ and grid spacing Δx and Δy. Initialize the grid with the initial concentration field $C(\boldsymbol{x}, 0)$ and corrosion state $S(\boldsymbol{x}, 0)$ (1 for active corrosion sites, 0 otherwise). Set appropriate boundary conditions.

Implement Fick's Second Law:
Model the diffusion of corrosive species using the central finite difference scheme:

$$C^{n+1}[i,j] = C^n[i,j] + D \cdot \Delta t \left(\frac{C^n[i+1,j] - 2 \cdot C^n[i,j] + C^n[i-1,j]}{\Delta x^2} + \frac{C^n[i,j+1] - 2 \cdot C^n[i,j] + C^n[i,j-1]}{\Delta y^2} \right), \quad (4.2)$$

Incorporate Reaction Kinetics:
For each grid point, calculate the local potential $V(i,j)^n$ based on the concentration $C(i,j)^n$. Implement reaction kinetics:
- If $V(i,j)^n > V_{th}$ and $S(i,j)^n = 1$, update concentration due to anodic reaction:

$$C(i,j)^{n+1} = C(i,j)^{n+1} - k_a \Delta t$$

- If $V(i,j)^n < V_{th}$ and $S(i,j)^n = 1$, update concentration due to cathodic reaction:

$$C(i,j)^{n+1} = C(i,j)^{n+1} + k_c \Delta t$$

Cellular Automaton Update:
Update the corrosion state $S(i,j)^{n+1}$ based on local concentration $C(i,j)^{n+1}$ and neighboring states. Activate corrosion at a site if $C(i,j)^{n+1} > C_{crit}$ or if any neighboring site is corroded.

Apply Boundary Conditions:
Implement appropriate boundary conditions (e.g., Dirichlet or Neumann) for both the concentration field $C(\boldsymbol{x}, t)$ and the corrosion state $S(\boldsymbol{x}, t)$.

Simulation Loop:
Create a time-stepping loop to update concentration and corrosion state fields, applying diffusion, reaction kinetics, and CA updates.

Fig. 4 Prompt to simulate corrosion in metals with CA

Output:
Generate an animation plot showing the progression of corrosion over time. Optionally, save the animation as a GIF file.

Example Usage:

simulate_corrosion(Nx, Ny, D, initial_concentration, time_steps, Δt, Δx, Δy, k_a, k_c, V_{th})

Expected Output:
An animation plot of corrosion evolution with an option to save as a GIF file.

Fig. 4 (continued)

3 Code Listing

```
# To find a better animation, please use the saved gif file
import numpy as np
import matplotlib.pyplot as plt
import matplotlib.animation as animation
import matplotlib
from matplotlib.colors import ListedColormap
# Set the backend to 'TkAgg' for interactive plotting with
    ↪ Tkinter
matplotlib.use('TkAgg')
# Define the domain and initialize parameters with specific
    ↪ initial damage and corroded cells
def initialize_domain(N_x, N_y, protection_layer_thickness):
    C = np.zeros((N_x, N_y))
    # Set initial damage cells for the top crack
    top_crack_indices = [(slice(0, 2), slice(95, 105)),
                         (slice(3, 5), slice(96, 104)),
                         (slice(6, 8), slice(97, 103)),
                         (slice(9, 11), slice(99, 101)),
                         (slice(12, 14), slice(95, 101)),
                         (slice(15, 17), slice(94, 100)),]
    # Set initial damage cells for the bottom crack
    bottom_crack_indices = [(slice(198, 200), slice(95, 105)),
                            (slice(195, 197), slice(96, 104)),
                            (slice(192, 194), slice(97, 103)),
                            (slice(189, 191), slice(96, 102)),
                            (slice(186, 188), slice(95, 101)),
                            (slice(183, 185), slice(94, 100)),]
```

```
    for idx in top_crack_indices + bottom_crack_indices:
        C[idx] = -0.75  # Initial damage (black)
    # Add protective layer
    C[:protection_layer_thickness, :] = -1
    C[-protection_layer_thickness:, :] = -1
    C[:, :protection_layer_thickness] = -1
    C[:, -protection_layer_thickness:] = -1
    return C
# Implement Fick's second law for diffusion with different
    ↪ coefficients for each metal
def diffuse(C, D_top, D_bottom, delta_t, delta_x, delta_y):
    C_new = C.copy()
    random_factor = np.random.uniform(0.9, 1.1, size=C.shape)
    # Apply different diffusion coefficients based on the metal
    ↪ type
    for i in range(1, C.shape[0] - 1):
        for j in range(1, C.shape[1] - 1):
            if C[i, j] != -1 and C[i, j] != -0.75:  # Skip
    ↪ protective layer and initial damage
                if i < C.shape[0] // 2:
                    D = D_top
                else:
                    D = D_bottom
                C_new[i, j] += D * delta_t * ((C[i+1, j] - 2 * C[
    ↪ i, j] + C[i-1, j]) / delta_x ** 2 + (C[i, j+1] - 2 * C[i,
    ↪ j] + C[i, j-1]) / delta_y ** 2) * random_factor[i, j]
    return C_new
# Incorporate reaction kinetics with different corrosion rates
    ↪ for different metals
def apply_reaction_kinetics(C, state, reaction_threshold,
    ↪ corrosion_probability_top, corrosion_probability_low, step
    ↪ ):
    new_state = state.copy()
    if step < 50:
        return new_state  # No corrosion until step 50

    corroded = (C >= reaction_threshold) & (state == 0)
    # Determine the metal type for each cell
    is_top_metal = np.arange(state.shape[0]).reshape(-1, 1) <
    ↪ state.shape[0] // 2
    is_low_metal = ~is_top_metal
    # Apply corrosion probabilities
    corrosion_probability = np.where(is_top_metal,
    ↪ corrosion_probability_top, corrosion_probability_low)
    probabilistic_corrosion = np.random.rand(*C.shape) <
    ↪ corrosion_probability
    # Only corrode if at least one neighbor is corroded
    for i in range(1, state.shape[0] - 1):
        for j in range(1, state.shape[1] - 1):
            if corroded[i, j] and np.any(state[i-1:i+2, j-1:j+2]
    ↪ == 1):
                if probabilistic_corrosion[i, j]:
```

```
                    new_state[i, j] = 1  # New cells become
    ↪ corroded with specified probability
    return new_state
# Simulation function
def simulate_corrosion(N_x, N_y, D_top, D_bottom, time_steps,
    ↪ delta_x, delta_y, delta_t, protection_layer_thickness,
                        reaction_threshold,
    ↪ corrosion_probability_top, corrosion_probability_low,
    ↪ filename):
    C = initialize_domain(N_x, N_y, protection_layer_thickness)
    state = np.zeros_like(C)
    # Set initial corroded cells in the state array
    initial_corroded_indices = [(3, N_y // 2),(N_x - 4, N_y // 2)
    ↪ ]
    for idx in initial_corroded_indices:
        state[idx] = 1  # Initial corroded cells
    fig, (ax1, ax2) = plt.subplots(2, 1, figsize=(8, 10))  # Keep
    ↪  the figure size the same
    corroded_counts = []
    top_metal_counts = []
    bottom_metal_counts = []
    protective_layer_counts = []
    def update(t):
        nonlocal C, state
        if t >= 50:
            C = diffuse(C, D_top, D_bottom, delta_t, delta_x,
    ↪ delta_y)
        state = apply_reaction_kinetics(C, state,
    ↪ reaction_threshold, corrosion_probability_top,
    ↪ corrosion_probability_low, t)
        # Display state: corroded cells are marked with 1,
    ↪ initial damage with -0.75, protective layer with -1,
        # and different colors for different metals
        display_state = np.zeros_like(state, dtype=float)
        display_state[(state == 0) & (np.arange(state.shape[0]).
    ↪ reshape(-1, 1) < state.shape[0] // 2)] = -0.75  # Upper
    ↪ metal
        display_state[(state == 0) & (np.arange(state.shape[0]).
    ↪ reshape(-1, 1) >= state.shape[0] // 2)] = -0.25  # Lower
    ↪ metal
        display_state[C == -1] = -1  # Protective layer
        display_state[C == -0.75] = 0.75  # Initial damage
        display_state[state == 1] = 1  # Corroded cells
        ax1.clear()
        im = ax1.imshow(display_state, animated=True, cmap='
    ↪ inferno', vmin=-1, vmax=1)
        ax1.set_title(f'Time Step: {t + 1}')
        ax1.set_xlabel('X Coordinate')
        ax1.set_ylabel('Y Coordinate')
        # Custom legend
        colors = ['yellow', 'gold', 'midnightblue', 'darkviolet',
    ↪  'black']
```

```
        labels = ['Corroded Cells', 'Initial Damage', 'Top Metal'
    ↪ , 'Bottom Metal', 'Protective Layer']
        handles = [plt.Line2D([0], [0], marker='o', color='w',
    ↪ markerfacecolor=c, markersize=10) for c in colors]
        ax1.legend(handles, labels, bbox_to_anchor=(1.28, 1.05),
    ↪ loc='upper left', borderaxespad=0.)
        # Count the number of each type of cell
        corroded_count = np.sum(state == 1)
        top_metal_count = np.sum((state == 0) & (np.arange(state.
    ↪ shape[0]).reshape(-1, 1) < state.shape[0] // 2))
        bottom_metal_count = np.sum((state == 0) & (np.arange(
    ↪ state.shape[0]).reshape(-1, 1) >= state.shape[0] // 2))
        protective_layer_count = np.sum(C == -1)
        corroded_counts.append(corroded_count)
        top_metal_counts.append(top_metal_count)
        bottom_metal_counts.append(bottom_metal_count)
        protective_layer_counts.append(protective_layer_count)
        ax2.clear()
        ax2.plot(corroded_counts, color='yellow', label='Corroded
    ↪  Cells')
        ax2.plot(top_metal_counts, color='midnightblue', label='
    ↪ Top Metal')
        ax2.plot(bottom_metal_counts, color='darkviolet', label='
    ↪ Bottom Metal')
        ax2.plot(protective_layer_counts, color='black', label='
    ↪ Protective Layer')
        ax2.set_xlabel('Time Step')
        ax2.set_ylabel('Number of Cells')
        ax2.legend()
        return im,
    ani = animation.FuncAnimation(fig, update, frames=time_steps,
    ↪  repeat=False, interval=50, blit=True)
    # Set the colorbar
    cbar = plt.colorbar(ax1.imshow(np.zeros((N_x, N_y)), animated
    ↪ =True, cmap='inferno', vmin=-1, vmax=1), ax=ax1,
    ↪ orientation='vertical', pad=0.1)
    cbar.ax.set_ylabel('Corrosion Status')
    plt.subplots_adjust(left=0.2, right=0.85, top=0.85, bottom
    ↪ =0.2)  # Adjust plot size
    plt.tight_layout()  # Adjust layout to make space for legend
    # Save the animation as a GIF using Pillow
    ani.save(filename, writer='pillow')
    plt.show()
    return ani
# Example usage with specified initial conditions
N_x, N_y = 200, 200  # Dimensions of the plate
D_top = 1.4  # Diffusion coefficient for the top metal
D_bottom = 1.35  # Diffusion coefficient for the bottom metal
time_steps = 500  # Number of time steps to ensure full corrosion
delta_x = delta_y = 1  # Grid spacing
delta_t = 0.2  # Increased time step to ensure full corrosion
protection_layer_thickness = 2  # Thickness of the protective
    ↪ layer
```

```
reaction_threshold = 0.1  # Threshold concentration to trigger
    ↪ corrosion
# Corrosion probabilities
corrosion_probability_top = 0.35  # 35% chance of each cell being
    ↪  corroded in the top metal
corrosion_probability_low = 0.33  # 33% chance of each cell being
    ↪  corroded in the low metal
# Call the simulation function and save the animation
filename = 'corrosion_simulation.gif'
ani = simulate_corrosion(N_x, N_y, D_top, D_bottom, time_steps,
    ↪ delta_x, delta_y, delta_t, protection_layer_thickness,
                                reaction_threshold,
    ↪ corrosion_probability_top, corrosion_probability_low,
    ↪ filename)
```

Code Listing 4.1 Output by ChatGPT-4o for solving 2D Modeling Corrosion Using CA

4 Verification

This section verifies the implementation of FSL in the corrosion simulation code. Since the numerical solution of the Initial Boundary Value Problem (IBVP) is carried out by a explicit finite difference (FD) method, the verification implies a test of the FD scheme, and for the explicit character a check of the stability conditions, and conducting unit tests.

• **Finite Difference Scheme Verification.** Fick's second law in two dimensions, as given in

$$\frac{\partial C(x,y,t)}{\partial t} = D\left(\frac{\partial^2 C(x,y,t)}{\partial x^2} + \frac{\partial^2 C(x,y,t)}{\partial y^2}\right) \tag{2}$$

is approximated in the simulation code using a central finite difference scheme as shown in (2).

In this scheme, $C^n[i, j]$ represents the concentration at grid point (i, j) at time t_n, and $C^{n+1}[i, j]$ represents the concentration at time t_{n+1}. The terms $C[i+1, j]$, $C[i-1, j]$, $C[i, j+1]$, and $C[i, j-1]$ correspond to the concentrations at the neighboring grid points. Additionally, Δx and Δy are the grid spacings in the x and y directions, respectively, Δt is the time step, and D is the diffusion coefficient.

• **Stability Condition.** The stability condition for an explicit finite difference method in two dimensions is:

$$\Delta t \leq \frac{\Delta x^2}{2D} \tag{3}$$

Given the parameters $\Delta \boldsymbol{x} = \Delta \boldsymbol{y} = 1$ and the diffusion coefficients of $D_{\text{top}} = 1.4$ and $D_{\text{bottom}} = 1.35$, the stability limits are obtained as $\Delta t \leq \frac{1^2}{2 \cdot 1.4} \approx 0.357$ and $\Delta t \leq \frac{1^2}{2 \cdot 1.35} \approx 0.370$ in the x- and y-directions. With Δt set to 0.2 in the simulations, the stability condition is fulfilled.

• **Unit Tests Based on Specific Diffusion Function:** The verification code, as provided in Code Listing 4.2, is generated using ChatGPT-4o based on the algorithm outlined in Algorithm Box 4. In these tests, a simplified version of the diffusion function is used, excluding the random factor for clarity.

Algorithm 4: Calculation of Diffusion to Verify the Simulation

---

1 **Input:**

- C: 2D array representing the initial concentration field, with shape (Nx, Ny)
- D: Diffusion coefficient (scalar)
- Δt: Time step size (scalar)
- Δx, Δy: Grid spacing in the x and y directions, respectively (scalars)

Output:

- C_{n+1}: 2D array representing the updated concentration field after one time step

Steps:

1. Initialize $C_{n+1} \leftarrow C$ *(Create a copy of the initial concentration array).*
2. **For** each interior grid point (i, j) where $1 \leq i \leq Nx - 2$ and $1 \leq j \leq Ny - 2$:

   a. **If** $C[i, j] \neq -1$ and $C[i, j] \neq -0.75$ **then**:

      i. Update $C_{n+1}[i, j]$ using the following formula:

$$C(i, j)^{n+1} = C(i, j)^n + D\Delta t \left(\frac{C(i+1, j)^n - 2C(i, j)^n + C(i-1, j)^n}{\Delta x^2} + \frac{C(i, j+1)^n - 2C(i, j)^n + C(i, j-1)^n}{\Delta y^2} \right)$$

   b. **End If**.

3. **End For**.
4. Return C_{n+1} *(Return the updated concentration field).*

```
import numpy as np
def diffuse_test(C, D, delta_t, delta_x, delta_y):
    C_n_plus_1 = C.copy()
    for i in range(1, C.shape[0] - 1):
        for j in range(1, C.shape[1] - 1):
            if C[i, j] != -1 and C[i, j] != -0.75:
                C_new[i, j] += D * delta_t * ((C[i+1, j] - 2 * C[
    ↪ i, j] + C[i-1, j]) / delta_x ** 2 + (C[i, j+1] - 2 * C[i,
    ↪ j] + C[i, j-1]) / delta_y ** 2)
    return C_n_plus_1

def test_diffusion():
    C = np.zeros((5, 5))
    C[2, 2] = 1  # Initial concentration at the center
    D = 1.0
    delta_t = 0.2  # Time step for stability
    delta_x = delta_y = 1  # Grid spacing
    C_n_plus_1 = diffuse_test(C, D, delta_t, delta_x, delta_y)
    expected_center = 1 - 4 * D * delta_t / delta_x ** 2
    expected_adjacent = D * delta_t / delta_x ** 2
    assert np.isclose(C_new[2, 2], expected_center), f"Central
    ↪ value should decrease to {expected_center}"
    assert np.isclose(C_new[1, 2], expected_adjacent), f"Adjacent
    ↪  cells should increase to {expected_adjacent}"

test_diffusion()
print("Test passed!")
```

Code Listing 4.2 This code is used to ensure the implementation of FSL

In conclusion, the verification confirms that the diffusion implementation in the corrosion simulation code adheres to FSL. The finite difference scheme is correctly applied, the stability condition is met, and unit tests validate the expected diffusion behavior. Therefore, the diffusion component of the simulation is accurately implemented.

Reaction Kinetics

This section verifies that the implemented RK in the corrosion simulation obeys the expected behavior based on given parameters, ensuring the corrosion process follows the reaction threshold and corrosion probabilities for different metal types.

The code used for verification, as detailed in Code Listing 4.3, is generated by ChatGPT-4o following the algorithm presented in Algorithm Box 5. A test function initializes a grid representing a metal plate with specific initial conditions for both corrosion and concentration. The reaction kinetics function is then applied to verify if the cells corrode as expected. To ensure consistent results, the random seed is fixed, and specific cells are deliberately corroded.

Algorithm 5: Calculation of Reaction Kinetics to Verify the Corrosion Modeling

1 **Input:**

- C: 2D array representing the concentration field, with shape (Nx, Ny)
- $state$: 2D array representing the current corrosion state, with shape (Nx, Ny)
- $reaction\_threshold$: Scalar value representing the concentration threshold for corrosion
- $corrosion\_probability\_top$: Probability of corrosion for the top metal layer
- $corrosion\_probability\_low$: Probability of corrosion for the bottom metal layer
- $step$: Integer representing the current simulation step

Output:

- new_state: 2D array representing the updated corrosion state

Procedure:

1. Initialize $new\_state \leftarrow state$.
2. **If** $step < 50$, **then** return new_state.
3. Identify cells where $C \geq reaction\_threshold$ and $state == 0$.
4. Create a boolean mask is_top_metal to identify top metal regions.
5. Assign $corrosion\_probability$ based on is_top_metal.
6. Set a fixed random seed for reproducibility.
7. Generate a random array $probabilistic\_corrosion$.
8. Manually set $probabilistic\_corrosion[4, 5]$ and $probabilistic\_corrosion[6, 5]$ to 0.
9. **For** each interior grid point (i, j):

   a. **If** $corroded[i, j]$ is True:
      i. Extract neighbors of $state[i, j]$.
      ii. **If** any neighbor is corroded:
         A. **If** $probabilistic\_corrosion[i, j] < corrosion\_probability[i, j]$, **then** set $new\_state[i, j] = 1$.
      iii. **Else**, print a message indicating no corrosion.
   b. **Else**, print a message indicating no corroded neighbors.

10. Return new_state.

```python
import numpy as np

# Corrected apply_reaction_kinetics function with deterministic
    ↪ outcome for testing
def apply_reaction_kinetics(C, state, reaction_threshold,
    ↪ corrosion_probability_top, corrosion_probability_low, step
    ↪ ):
    new_state = state.copy()
    if step < 50:
        print(f"Step {step}: No corrosion applied (initial delay)
    ↪ .")
        return new_state  # No corrosion until step 50

    corroded = (C >= reaction_threshold) & (state == 0)
    is_top_metal = np.zeros_like(C, dtype=bool)
    is_top_metal[:C.shape[0] // 2, :] = True  # First half rows
    ↪ are top metal
    corrosion_probability = np.where(is_top_metal,
    ↪ corrosion_probability_top, corrosion_probability_low)
```

```
    # Set a fixed random seed for reproducibility
    np.random.seed(0)
    probabilistic_corrosion = np.random.rand(*C.shape)
    # Force the cells [4, 5] and [6, 5] to corrode for testing
    ↪ purposes
    probabilistic_corrosion[4, 5] = 0
    probabilistic_corrosion[6, 5] = 0

    for i in range(1, state.shape[0] - 1):
        for j in range(1, state.shape[1] - 1):
            if corroded[i, j]:
                neighbors = state[i - 1:i + 2, j - 1:j + 2]
                if np.any(neighbors == 1):
                    if probabilistic_corrosion[i, j] <
    ↪ corrosion_probability[i, j]:
                        new_state[i, j] = 1  # New cells become
    ↪ corroded with specified probability
                        print(
                            f"Cell ({i}, {j}) corroded. Neighbors
    ↪ : {neighbors.flatten()}, Prob: {corrosion_probability[i, j
    ↪ ]}, Rand: {probabilistic_corrosion[i, j]}")
                    else:
                        print(
                            f"Cell ({i}, {j}) not corroded.
    ↪ Neighbors: {neighbors.flatten()}, Prob: {
    ↪ corrosion_probability[i, j]}, Rand: {
    ↪ probabilistic_corrosion[i, j]}")
                else:
                    print(f"Cell ({i}, {j}) has no corroded
    ↪ neighbors.")
            else:
                print(f"Cell ({i}, {j}) not above reaction
    ↪ threshold.")
    return new_state

# Test function for reaction kinetics
# The test function initializes the grid and verifies the
    ↪ corrosion process.
def test_reaction_kinetics():
    N_x, N_y = 10, 10  # Small grid for testing
    C = np.zeros((N_x, N_y))
    state = np.zeros((N_x, N_y))
    # Set initial conditions
    reaction_threshold = 0.1
    corrosion_probability_top = 0.5
    corrosion_probability_low = 0.3
    # Set specific cells to meet the corrosion threshold
    C[4, 5] = 0.2  # Above threshold in top metal
    C[6, 5] = 0.2  # Above threshold in bottom metal
    state[4, 4] = 1  # Initial corroded cell near top threshold
    ↪ cell
```

```
    state[6, 6] = 1  # Initial corroded cell near bottom
    ↪ threshold cell

    # Print initial state and concentration
    print("Initial State:")
    print(state)
    print("Initial Concentration (C):")
    print(C)

    # Apply reaction kinetics
    step = 51  # Step beyond the initial delay
    new_state = apply_reaction_kinetics(C, state,
    ↪ reaction_threshold, corrosion_probability_top,
    ↪ corrosion_probability_low, step)

    # Print new state after applying reaction kinetics
    print("New State After Reaction Kinetics:")
    print(new_state)

    # Check if the specific cells have corroded based on their
    ↪ neighbors and probabilities
    assert new_state[4, 5] == 1, "Top metal cell should be
    ↪ corroded."
    assert new_state[6, 5] == 1, "Bottom metal cell should be
    ↪ corroded."

# Run the test
test_reaction_kinetics()
```

Code Listing 4.3 Code to test the correct implementation of reaction kinetics

4.1 Results

Initial conditions are set with specific cells at the corrosion threshold. The reaction kinetics function is applied, with debug statements providing insights into the process. As expected, the cells at [4, 5] and [6, 5] corrode, confirming the function's behavior under controlled conditions as underpinned in Fig. 5.

The verification process demonstrates that the RK implementation obeys the expected behavior based on given parameters. By using a fixed random seed and forcing specific cells to corrode, deterministic outcomes are ensured, confirming the implementation's correctness. This approach can be extended to validate other aspects of the corrosion simulation.

The code provided in the repository [6] is used to specify the initial and boundary conditions of the problem. The feasibility of the research is also ensured by verifying that the obtained results closely match with a high degree of accuracy.

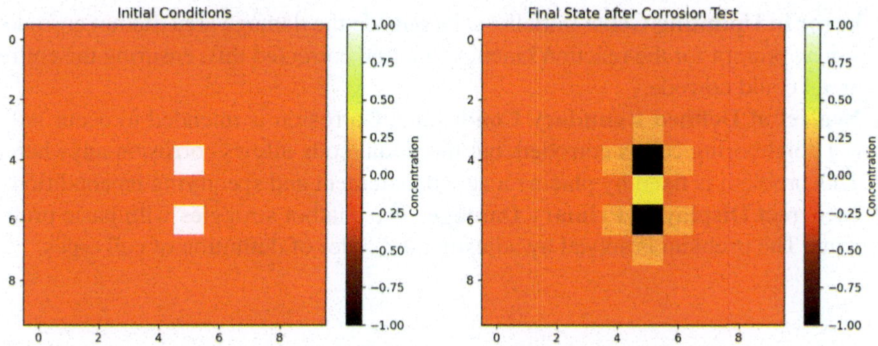

(a) Initial concentration matrix with cells [4, 5] and [6, 5] meeting the corrosion threshold

(b) Final matrix after corrosion. Cells [4, 5] and [6, 5] corroded

Fig. 5 Comparison of initial and final concentration matrices in the RK test

5 Discussion

Repeatability as a Reliability Signature The main form of the chatbot's responses remains consistent, with similar reactions applicable to similar prompts. However, there are definite differences in the details.

5.1 Errors, Omissions

The shortcomings can be summarized as follows:

- **Rule Violations**: The code generated by ChatGPT-4o sometimes violates basic rules. For example, corroded parts of the material are allowed to revert to their original state, which is unrealistic.
- **Inability to Fix Visualization Errors**: Some visualization errors in the generated code required manual intervention despite attempts to resolve them with secondary prompts.
- **Improper Plot Labeling**: Some labels in the legend section are incorrectly printed on top of each other, resolved by adjusting the plot size.
- **Incorrect Implementation of Fick's second law**: The initial simulation did not correctly follow FSL. This is resolved with a secondary prompt using the appropriate function.
- **Inadequate Initial Conditions**: The generated code displays initial damage incorrectly and contains errors, which are resolved by manual code editing. In this edition, the coordinates of the cells that suffered initial damage are accurately entered.

- **Error in Updating Cells**: Initially, cells outside the damage area did not corrode. A new prompt for the reaction kinetics function corrected this, ensuring the entire metal could corrode.
- **Neglect of Defined Boundary Conditions**: Corrosion is intended to occur only if a neighboring cell is corroded, but the initial code allows corrosion anywhere. This problem is fixed by placing a new if statement and specifying its condition.
- **Incorrect Display of Primary Damage**: The chatbot struggles with basic problems. The problem is solved by adjusting the order of definition of cell types.

5.2 Learned Lessons

In conclusion, ChatGPT-4o is a powerful research assistant capable of significantly streamlining the research process by providing comprehensive information, data analysis, and insights. However, the results generated by ChatGPT-4o require expert review to correct any minor issues or inaccuracies, ensuring the reliability and accuracy of the information. The chatbot's effectiveness is highly dependent on proper utilization by the user, who must understand its strengths and limitations to maximize response quality. When used correctly, ChatGPT can deliver highly valuable results, contributing meaningfully to research endeavors.

The project highlights the substantial improvements in the latest iteration of ChatGPT, specifically the difference between versions 3.5 and 4o. ChatGPT-4o exhibits a marked enhancement in contextual understanding, information retrieval, and response coherence, making it a more effective and reliable research tool. These advancements enable users to tackle more complex tasks, saving time and improving output quality. While expert review remains crucial, the enhanced capabilities of ChatGPT-4o significantly boost its utility as an intelligent and dependable research assistant.

5.3 Refined Corrosion Simulation Prompt

The prompt shown in Fig. 6 is used to edit the code generated by the initial prompt shown in Fig. 4. The final code is shown in Code Listing 4.1.

Objective:

Enhance the initial Python program for 2D corrosion simulation by incorporating specific initial conditions and domain modifications.

Instructions:

- **Initial and Boundary Conditions**:
  - Place the initial damage in the form of two cracks at the top and bottom of each metal.
  - Corrosive substances should enter through these areas.
- **Corrosion Propagation**:
  - Corrosion should begin at the initial damage sites and spread throughout the metal.
- **Protective Layer**:
  - Add a protective layer around the part to inhibit corrosion.
  - Use different colors to visually represent the protective layer.
- **Domain Specifications**:
  - Set the plot dimensions to 200 x 200 cells.
- **State Change Rules**:
  - Once a cell becomes corroded, it cannot revert to its previous state and remains corroded for the duration of the simulation.
- **Material Heterogeneity**:
  - Modify the lower half of the part to be a different type of metal (represented by a different color). Set different values for each metal in diffusion coefficients and corrosion probabilities.

Comments:

Update the code to include these changes and add comments to explain the new features and modifications.

Expected Output:

An improved animation plot demonstrating the progression of corrosion over time, showcasing the effects of the protective layer and the varying corrosion rates of the different metals.

Fig. 6 Prompt used for the final editing of the code

Example Usage:
simulate_corrosion(N_x, N_y, D, initial_concentration, time_steps)

Expected Output:
An animation plot of corrosion evolution, with an option to save as a gif file.

Fig. 6 (continued)

5.4 Visual Results

This section presents the visual documentation of the corrosion simulation process at different stages highlighting in Fig. 7(a) the initial, in (b) an intermediate, in (c) an advanced and in (d) the final stage of the corrosion progress. Faster progress in the upper part of the simulation domain through a larger diffusion coefficient (compare the values in Table 1) is clearly visible.

6 Conclusion

The objective of this study was to develop a robust 2D CA model for simulating corrosion processes, incorporating FSL of diffusion and RK. The simulation tool serves the purpose to predict the initiation and propagation of corrosion under various environmental conditions with the ultimate goal of an effective corrosion prevention and control strategies.

The main results of the study are summarized as follows:

Model Development: A 2D CA model was successfully developed to simulate corrosion. The model accurately captures the transport of corrosive species and the subsequent chemical reactions at the material's surface.

Diffusion Verification: The implementation of FSL was rigorously tested. The verification process confirmed that the model correctly simulates the diffusion process, adhering to theoretical expectations and stability conditions.

Reaction Kinetics Verification: The reaction kinetics were verified through controlled testing. The results show that the model accurately simulates the corrosion process, taking into account the different corrosion rates for various metals and the influence of initial conditions.

Simulation Results: The simulation effectively demonstrated the progression of corrosion, showcasing the protective layer's impact and the different corrosion rates of the metals. The refined prompts and code adjustments ensured that the simulation adhered to the specified conditions.

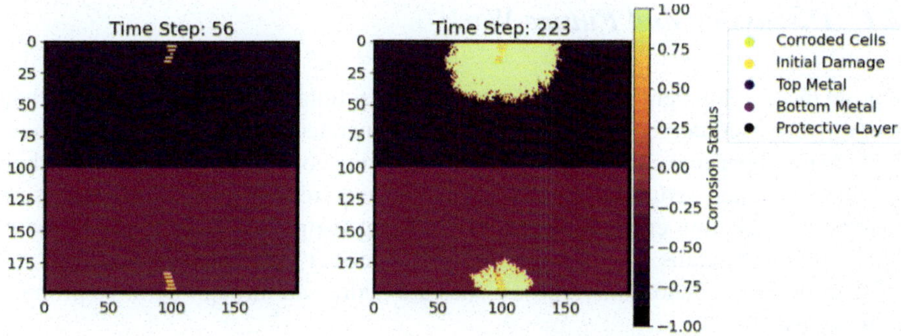

(a) Initial stage (b) Intermediate stage

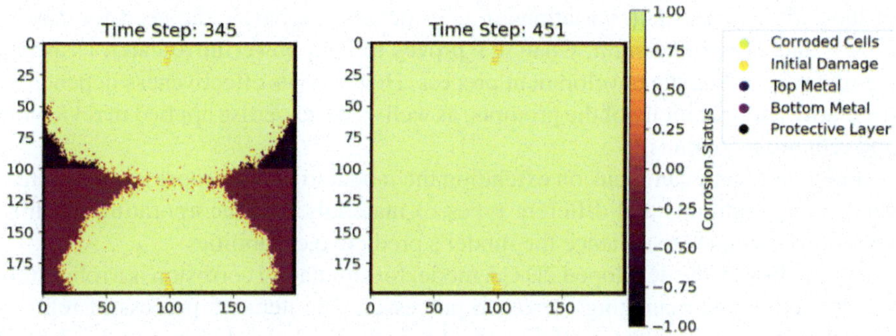

(c) Advanced stage (d) Final stage

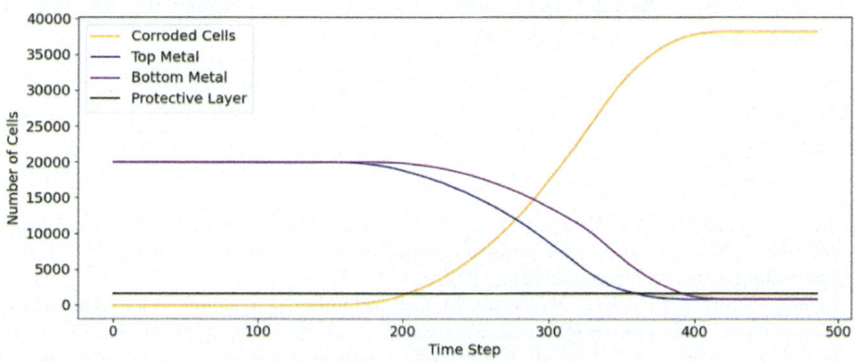

(e) Number of different cell types over time

Fig. 7 Different stages **a–d** of the corrosion process along with **e** temporal evolution of the different cell types

6.1 Discussion and Future Work

During the development of the 2D CA model for simulating corrosion, several challenges were encountered with the initial code generated by ChatGPT. Major issues include improper rule enforcement, such as allowing corroded cells to revert to their original state, and visualization challenges that require manual fixes. Minor issues involve plot labeling errors resolved by adjusting the plot size. The initial prompt lacks sufficient details, leading to issues like incorrect implementation of FSL and inadequate initial conditions. These issues are addressed through iterative prompts and manual edits.

Despite these challenges, the process highlights the importance of precise and detailed prompts to ensure the accuracy of the generated code. While the structure of the code is consistent, variations in finer details emphasize the need for careful review and expert oversight. ChatGPT proves to be a powerful research assistant, significantly aiding the development process. However, its effectiveness depends on the clarity and specificity of the prompts, as well as the expertise applied in reviewing and refining its outputs.

Future research can focus on extending the model to include more complex environmental conditions and different types of materials, and incorporating machine learning techniques to enhance the model's predictive capabilities.

In conclusion, the developed 2D CA model for simulating corrosion is a robust tool for predicting and managing corrosion processes. The iterative process of refining prompts and verifying the model ensured that the final simulation met the desired standards of accuracy and reliability. This study underscores the potential of AI-assisted tools like ChatGPT in scientific research, while emphasizing the need for continuous refinement and expert oversight.

References

1. Wang, Y., Guan, B., Wei, X., Lu, J., Ding, J. High-temperature corrosion of Ni-based alloys in chloride molten salts: A cellular automaton model. *Corrosion Science*. **150**, pp. 17-28 (2019). https://doi.org/10.1016/j.solmat.2019.110170.
2. Reinoso-Burrows, J.C., Toro, N., Cortés, M. Cellular automata modeling as a tool in corrosion management. *Materials*. **16**(6051) (2023). https://doi.org/10.3390/ma16176051.
3. Stafiej, J., di Caprio, D., Bartosik, L. Corrosion-passivation processes in a cellular automata-based simulation study. *Journal of Supercomputing*. **65**, pp. 697-709 (2013). https://doi.org/10.1007/s11227-013-0933-8.
4. Valor, A., Caleyo, F., Hallen, J.M. Pitting corrosion modeling in low-carbon steels using cellular automata. *Corrosion*. **66**(2), pp. 025004-1–025004-8 (2010).
5. Chen, H., Chen, Y., Zhang, J. Cellular automaton modeling on the corrosion/oxidation mechanism of steel in a liquid metal environment. *Progress in Nuclear Energy*. **50**(5-6), pp. 587–593 (2008). Elsevier. https://doi.org/10.1016/j.pnucene.2007.11.044.
6. Gooo, T. Cellular Automata Corrosion. (2021). Available at: https://github.com/tracygooo/cellular-automata-corrosion. Code posted: 21 June 2021, Accessed: 5 May 2024.

7. Chen, M.C., Wen, Q.Q. Simulation of corrosion process for structure with the cellular automata method. *IOP Conference Series: Materials Science and Engineering*. **216**, 012012 (2017). IOP Publishing. https://doi.org/10.1088/1757-899X/216/1/012012.
8. Xiao, Z., Jun, H., Yuqi, W., Maosheng, Z., Zaoxiao, Z. Simulation of pitting corrosion for Ni-based alloy using a cellular automata model. *Rare Metal Materials and Engineering*. **44**(10), pp. 2347–2352 (2015). Elsevier. https://doi.org/10.1016/S1875-5372(16)30018-2

Instationary Heat Conduction on Rectangular Domains with Arbitrary Circular Holes

Aagashram Neelakandan and Bernhard Eidel

Abstract This chapter presents the generation and evaluation of Python code for simulating two-dimensional, unsteady heat conduction in rectangular domains with multiple holes using ChatGPT-4. The focus is on solving the Fourier heat conduction equation through the Forward Time Centered Space (FTCS) scheme, a finite difference method. The chapter begins with a detailed derivation of the algorithmic approach, transitioning from the governing partial differential equation to the discrete FTCS scheme for temperature updates. The accuracy and reliability of the generated code are rigorously tested, including a comparison with analytical solutions, demonstrating the effectiveness of AI-assisted coding in computational materials science and mechanics.

1 Introduction

The study focuses on understanding the performance and usage of ChatGPT-4's ability to generate code for a scientific problem by proposing the implementation of the 2D Fourier transient heat conduction problem using the Forward Time Centered Space (FTCS) scheme [5].

The heat conduction equation describes how heat is transferred within a material over time. In its general form, the 2D heat conduction equation is given by

$$\frac{\partial T}{\partial t} = \alpha \left(\frac{\partial^2 T}{\partial x^2} + \frac{\partial^2 T}{\partial y^2} \right) , \tag{1}$$

A. Neelakandan (✉) · B. Eidel
M5, IMFD, TUBAF, Freiberg, Germany
e-mail: aagashram.neelakandan@student.tu-freiberg.de

B. Eidel
e-mail: bernhard.eidel@imfd.tu-freiberg.de

© The Author(s), under exclusive license to Springer Nature Switzerland AG 2025
B. Eidel (ed.), *GPT for Python-Coding in Computational Materials Science and Mechanics*, Studies in Computational Intelligence 1198,
https://doi.org/10.1007/978-3-031-85470-5_5

where T represents the temperature, t is time, x and y are the spatial coordinates, and α is the thermal diffusivity. This equation represents the conservation of energy and relates the temporal change in temperature to the spatial variations and accounts for the thermal diffusivity of the material.

The FTCS scheme is a numerical method used to approximate the solutions of partial differential equations, which is based on the Finite Difference Method (FDM), having discretization for both space and time quantities. The FTCS approximates the spatial derivatives using the central differences scheme and time derivatives using the forward Euler method.

To apply the FTCS scheme, we discretize the spatial and temporal domains. Let Δx and Δy be the spatial step sizes in the x and y directions, respectively, and let Δt be the time step. Define the grid points as $x_i = i\Delta x$ and $y_j = j\Delta y$ for integers i and j, and $t_n = n\Delta t$ for integer n. The temperature at a grid point (x_i, y_j) at time t_n is denoted as $T_{i,j}^n$.

The finite difference approximations for the first and second derivatives are given by

$$\frac{\partial T}{\partial t} \approx \frac{T_{i,j}^{n+1} - T_{i,j}^n}{\Delta t}, \tag{2}$$

$$\frac{\partial^2 T}{\partial x^2} \approx \frac{T_{i+1,j}^n - 2T_{i,j}^n + T_{i-1,j}^n}{\Delta x^2}, \tag{3}$$

$$\frac{\partial^2 T}{\partial y^2} \approx \frac{T_{i,j+1}^n - 2T_{i,j}^n + T_{i,j-1}^n}{\Delta y^2}. \tag{4}$$

Substituting equations (2), (3), and (4) into the heat conduction equation (1), we obtain the FTCS scheme

$$T_{i,j}^{n+1} = T_{i,j}^n + \alpha\Delta t \left(\frac{T_{i+1,j}^n - 2T_{i,j}^n + T_{i-1,j}^n}{\Delta x^2} + \frac{T_{i,j+1}^n - 2T_{i,j}^n + T_{i,j-1}^n}{\Delta y^2} \right), \tag{5}$$

where $T_{i,j}^{n+1}$ is the temperature at time t_{n+1} calculated from known temperature $T_{i,j}^n$ at time t_n and $\Delta t = t_{n+1} - t_n$ is the discretized time step with Δx and Δy being the spatial step sizes as shown in Fig. 1.

The algorithm for instationary heat conduction is given in the algorithm box 1 and the prompt used for generating it is mentioned in the box Fig. 2.

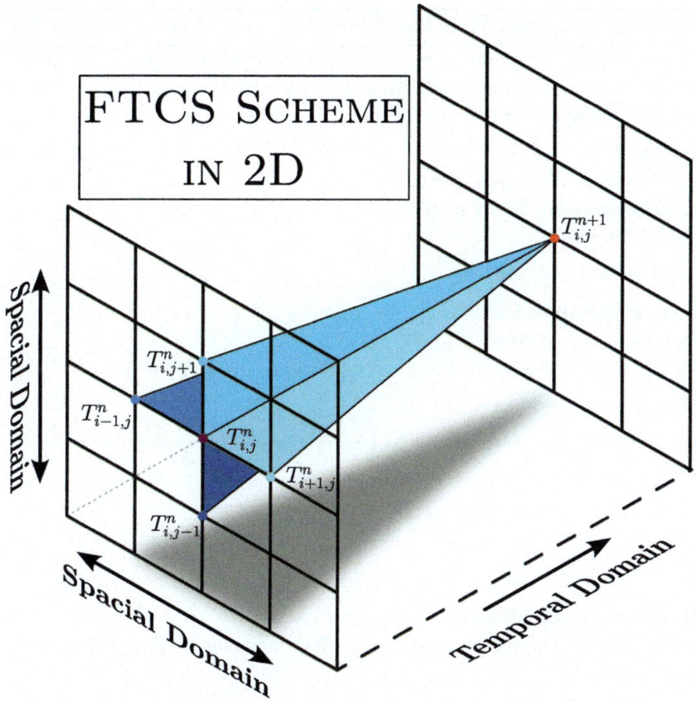

Fig. 1 2D representation of FTCS scheme based on FDM indicating the spacial points ($T_{i,j}^n$, $T_{i+1,j}^n$, $T_{i-1,j}^n$, $T_{i,j+1}^n$, $T_{i,j-1}^n$) used for calculating next temporal point ($T_{i,j}^{n+1}$)

> Generate the LaTeX code for an algorithm that solves a 2D transient heat conduction problem using the Forward Time Centered Space (FTCS) scheme. The algorithm should be formatted using the algorithm environment in LaTeX.

Fig. 2 Prompt for generating the algorithm in LaTeX pseudocode for instationary heat conduction using an FTCS scheme in 2D

Algorithm 1: Instationary Heat Conduction using FTCS Scheme in 2D

Input: Grid size (Nx, Ny), time steps T, time step Δt, grid spacing Δx, Δy, thermal diffusivity α, initial temperature field $T(\mathbf{x}, 0)$, boundary conditions
Output: Temperature field $T(\mathbf{x}, t)$ over time

```
Initialization:
Set the grid with size (Nx, Ny);
Initialize the temperature field T(x, 0);
Set thermal diffusivity α;
for t = 0toT with step Δt do
    Update Temperature Field:
    foreach interior grid point (i, j) do
        Calculate the temperature T(i, j)^{n+1} at the next time step using the FTCS scheme:
```

$$T(i, j)^{n+1} = T(i, j)^n + \alpha\Delta t \left(\frac{T(i+1, j)^n - 2T(i, j)^n + T(i-1, j)^n}{\Delta x^2} + \frac{T(i, j+1)^n - 2T(i, j)^n + T(i, j-1)^n}{\Delta y^2} \right)$$

```
    end
    Apply Boundary Conditions:
    Apply the specified boundary conditions to the temperature field T(x, t);
end
Output:
Return the temperature field T(x, t) over time;
```

2 Prompt

The prompt aims to generate code using ChatGPT-4 to numerically compute the time-dependent temperature field in a rectangular 2D domain with arbitrary dimensions under the constraints of an arbitrary number of different circular holes in the domain by applying the FTCS scheme mentioned before in the theory for solving the heat conduction equation.

For formulating a prompt for ChatGPT-4 to generate a 2D transient heat conduction simulation in Python using the FTCS method, we have to mention every critical parameter. These parameters encompass plate dimensions, thermal diffusivity, spatial step, and maximum iteration time. The prompt 5.2 outlines precise boundary conditions, incorporating Dirichlet at the top and Neumann on the other sides while describing hole configurations with coordinates, radius, type, and value. The prompt also includes defining functions for grid initialization, hole condition application, temperature distribution computation, and high-resolution visual output generation. Experiments with prompts reveal that some functions need heightened emphasis, such as hole evolution, visualization, and verbose logging. Verbose logging purpose is to provide extensive and detailed information about the program's status, often

I want you to act as scientific Python programmer who has expertise in mechanics, computational material science and thermodynamics. Your job is to provide a complete production ready working code and below is the prompt.

Prompt for Generating a 2D Heat Conduction Simulation Script in Python:

Create a Python script to simulate 2D transient heat conduction on a square plate using the Forward Time Centered Space (FTCS) method. The script should be pre-configured with numerical parameters and execute without requiring runtime user input, featuring detailed comments for educational purposes.

Required Features and Parameters:

- **Fixed Numerical Parameters:** Set `plate_length` as 50.0 units, `max_iter_time` as 750 iterations, alpha (thermal diffusivity) as 2.0, and `delta_x` (spatial step) as 1.0. Automatically calculate `delta_t` (time step) and gamma (FTCS constant) based on these values.
- **Boundary Conditions:** Implement a boundaries dictionary with default settings: Dirichlet condition of 100.0 at the top and Neumann conditions of 0.0 on other sides.
- **Hole Configuration:** In a holes list, define a default center hole with coordinates (25.0, 25.0), radius 5.0, Dirichlet type 'D', and value 0.0. The holes should be visually distinct in the simulation, and their thermal evolution should be accurately represented.

Fig. 3 Prompt or solving 2D Fourier transient heat conduction problem

used for debugging or monitoring purposes. The prompt also calls for library availability checks, prioritizing OpenCV for video creation and seamlessly falling back to Pillow if required. The anticipated outcome is a meticulously documented, self-contained script, poised to deliver precise simulations with explicit hole evolution visibility, archived as both a video and GIF (Fig. 3).

3 Generation Results

The reason that ChatGPT-4 cannot consistently generate the same results is due to its non-deterministic nature, influenced by random sampling during generation and contextual dependencies. The model's training on a diverse dataset contributes to response variability. To alleviate these deficiencies and speed up programming, one

- **FTCS Simulation Functions:** Write functions for initializing the grid with boundary conditions, applying hole configurations, computing the temperature distribution, and generating visual outputs with high-resolution heatmaps.
- **Verbose Output:** Throughout the simulation, provide a verbose output that shows the evolution of the holes, including temperature changes at each time step.
- **Output Resolution:** Ensure generated heatmaps are suitable for a 720P resolution output. If OpenCV is available, compile the heatmaps into a video at 1280x720 pixels; otherwise, create a high-definition GIF using Pillow.
- **Execution Flow:** Design the main function to systematically prepare the simulation environment, apply hole conditions, run the FTCS calculations, and save the visual results.

Explicit Instructions:

- **Visual Emphasis on Holes:** Ensure that holes remain clearly visible throughout the simulation, with their boundary conditions properly influencing the temperature distribution without being overwritten by the surrounding material's temperature.
- **Verbose Evolution Display:** Implement verbose logging to track and display the temperature evolution within the holes at each simulation step.
- **Library Availability Checks:** The script must check for the presence of OpenCV for video creation. If OpenCV is not available, fallback to Pillow should be enabled for GIF output.

Outcome Expectations:
The script should be self-contained, well-commented, and ready to run, producing accurate simulation results with the given default parameters. The final simulation should visibly and accurately show the evolution of the holes and save the results in both video (if OpenCV is present) and GIF formats for easy review.

Fig. 3 (continued)

needs to control randomness by refining input through elaborate prompt engineering and regenerating the answers, and the user should understand that complete generation is not guaranteed in a single prompt. The following code in code Listing 3 is generated by the ChatGPT-4 with the given prompt, without any additional inputs and one of the output plot is shown in Fig. 4.

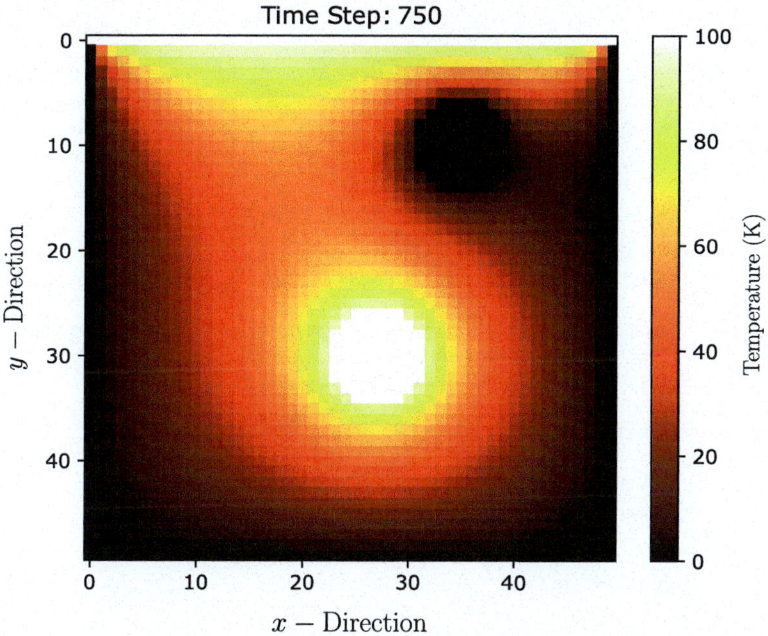

Fig. 4 Contour plot of the plate domain (quantities on the abscissa x and ordinate y in (mm)) with two circular holes at time step $t = 749$ s, with a time step size of $\Delta t = 0.125$ s. The top boundary is maintained at a temperature of 100 K, while the other boundaries are kept at 0 K. The first hole, centered at (10, 35), has a temperature of 0 K, and the second hole, centered at (30, 27), has a temperature of 100 K

```
import numpy as np
import matplotlib.pyplot as plt
import matplotlib.animation as animation
from PIL import Image
import os

# Check for OpenCV availability
try:
    import cv2
    OPENCV_AVAILABLE = True
except ImportError:
    OPENCV_AVAILABLE = False

# Fixed Numerical Parameters
plate_length = 50.0
max_iter_time = 750
alpha = 2.0
delta_x = 1.0
delta_t = (delta_x ** 2) / (4 * alpha)  # Stability
    ↪ criterion
gamma = alpha * delta_t / (delta_x ** 2)
```

```python

# Boundary Conditions
boundaries = {'top': 100.0, 'bottom': 0.0, 'left': 0.0, 'right': 0.0}

# Hole Configuration
holes = [{'center': (25, 25), 'radius': 5, 'type': 'D', 'value': 0.0}]

def initialize_grid():
    """Initializes the temperature grid with boundary conditions."""
    grid = np.zeros((int(plate_length), int(plate_length)))
    grid[0, :] = boundaries['top']
    # Neumann conditions on other sides are implicitly set to zero
    return grid

def apply_hole_conditions(grid):
    """Applies hole configurations to the grid."""
    for hole in holes:
        x_center, y_center = hole['center']
        radius = hole['radius']
        for i in range(grid.shape[0]):
            for j in range(grid.shape[1]):
                if (i - x_center) ** 2 + (j - y_center) ** 2 < radius ** 2:
                    grid[i, j] = hole['value']
    return grid

def FTCS(grid):
    """Performs a single time step of the FTCS method."""
    new_grid = np.copy(grid)
    for i in range(1, grid.shape[0] - 1):
        for j in range(1, grid.shape[1] - 1):
            new_grid[i, j] = grid[i, j] + gamma * (grid[i+1, j] + grid[i-1, j] + grid[i, j+1] + grid[i, j-1] - 4 * grid[i, j])
    return apply_hole_conditions(new_grid)

def save_frame(frame, iter_time, frames_dir):
    """Saves a single frame of the simulation."""
    plt.imshow(frame, cmap='hot', interpolation='nearest')
    plt.title(f'Time Step: {iter_time}')
    plt.colorbar()
    plt.savefig(f"{frames_dir}/frame_{iter_time:04d}.png")
    plt.close()

def compile_video(frames_dir, output_file):
    """Compiles frames into a video."""
    if not OPENCV_AVAILABLE:
        return
    images = [img for img in os.listdir(frames_dir) if img.endswith(".png")]
    frame = cv2.imread(os.path.join(frames_dir, images[0]))
```

```python
    height, width, layers = frame.shape
    video = cv2.VideoWriter(output_file, cv2.VideoWriter_fourcc(*'DIVX'), 15, (width, height))
    for image in images:
        video.write(cv2.imread(os.path.join(frames_dir, image)))
    cv2.destroyAllWindows()
    video.release()

def compile_gif(frames_dir, output_file):
    """Compiles frames into a GIF."""
    images = [Image.open(os.path.join(frames_dir, img)) for img in os.listdir(frames_dir) if img.endswith(".png")]
    images[0].save(output_file, save_all=True, append_images=images[1:], optimize=False, duration=40, loop=0)

def run_simulation():
    """Runs the entire simulation."""
    grid = initialize_grid()
    grid = apply_hole_conditions(grid)
    frames_dir = 'frames'
    os.makedirs(frames_dir, exist_ok=True)

    for iter_time in range(max_iter_time):
        print(f"Iteration {iter_time}/{max_iter_time}")
        save_frame(grid, iter_time, frames_dir)
        grid = FTCS(grid)

    if OPENCV_AVAILABLE:
        compile_video(frames_dir, 'heat_conduction_simulation.avi')
    else:
        compile_gif(frames_dir, 'heat_conduction_simulation.gif')

if __name__ == "__main__":
    run_simulation()
```

Code Listing 5.1 Output by ChatGPT for solving 2D Fourier transient heat conduction problem

4 Testing and Verification

To assess ChatGPT-4's code generation accuracy, we employ intrinsic verification [3]. This involves validating the output by comparing it to established benchmarks, such as the one-dimensional heat conduction problem. For this validation, we set the left boundary of the plate ($x = 0$ mm) at a temperature $T_1 = 1000$ K and the right side ($x = 1000$ mm) at a lower temperature of $T_2 = 0$ K, creating a temperature gradient in the x-direction. This reduces the problem to one-dimensional, for which the closed form could be obtained by solving (1) in 1D, which yields (6) expressed in terms of the error function erf

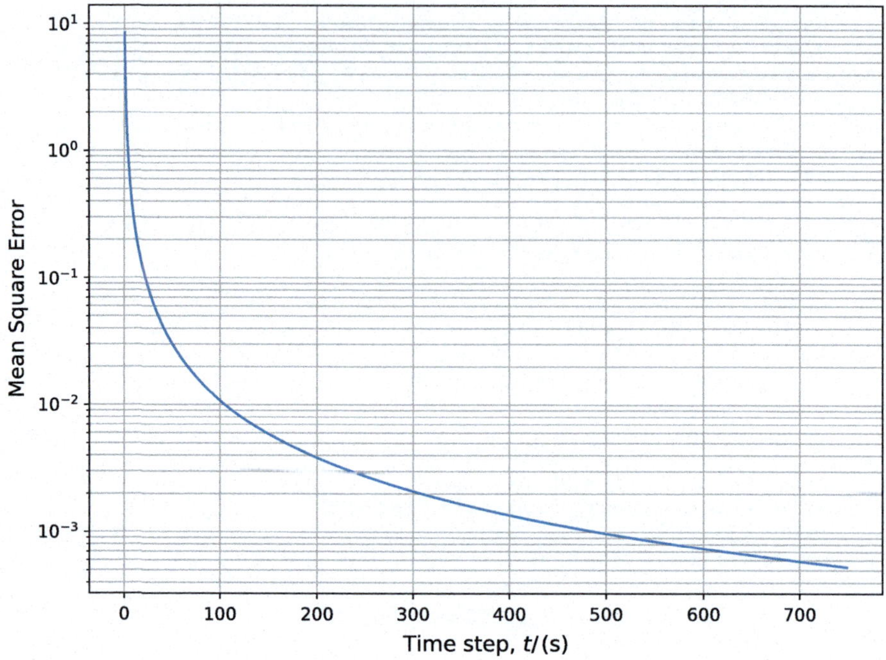

Fig. 5 MSE plotted in logarithmic scale between FTCS and Analytical Solutions over time steps along the middle of the plate

$$T(x,t) = T_1 - (T_1 - T_2) * \operatorname{erf}\left(\frac{x}{2\sqrt{\alpha t}}\right), \tag{6}$$

$$\text{where} \quad \operatorname{erf} = \frac{2}{\sqrt{\pi}} \int_0^z e^{-y^2} dy\,. \tag{7}$$

The analytical solution serves to verify the code-generated numerical solution, where the deviation is measured by the Mean Square Error (MSE) defined as

$$\text{MSE} = \frac{1}{n} \sum_{i=1}^{n} (T_{\text{analytical},i} - T_{\text{FTCS},i})^2\,, \tag{8}$$

where n is the number of data points along the plate's midsection, $T_{\text{analytical},i}$ is the temperature at point i computed using the analytical solution and $T_{\text{FTCS},i}$ is the temperature at point i computed using the FTCS scheme.

The diagrams in Figs. 5 and 6 illustrate the MSE between temperatures computed by the analytical and FTCS scheme along the plate's midsection. The reducing error over extended time intervals, coupled with the overall error bounds within 10% for a 1000 mm plate, indicates the use of the code generated by ChatGPT-4 directly, with or without minor adjustments for solving a scientific task is a viable option.

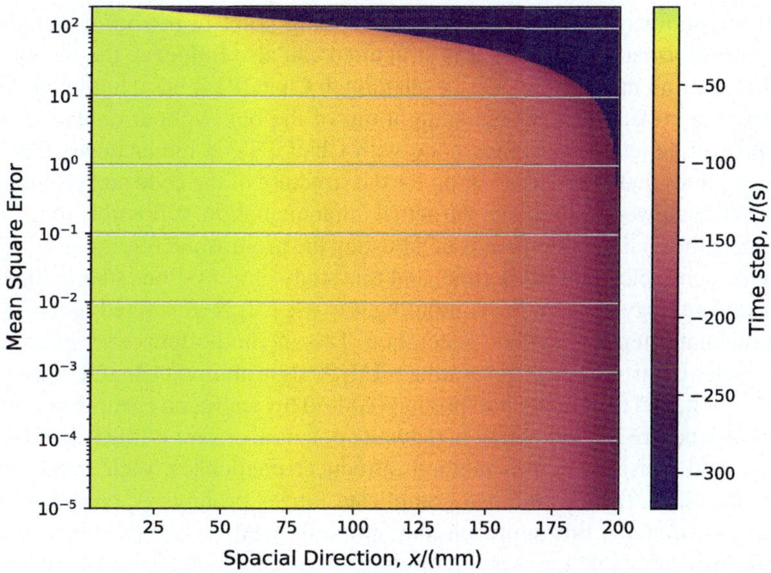

Fig. 6 Contour plot of the logarithm of MSE between FTCS and the analytical solutions over the spatial distance x in the middle of the plate at $y = 500$ mm, displayed up to a distance of $x \leq 200$ mm. The contour levels are colored to indicate the time steps of the simulation

5 Discussion

The output of the code generated by the ChatGPT-4 and its verification paves the strong foundation for using the Large Language Models (LLMs) as a critical tool in solving scientific problems. Even though they have issues with faithfulness in generated data caused by the sources of data, training, and inference, also known as *hallucinations* [2], which may result in outputs that are factually incorrect or flawed in reasoning due to memorized data, various methods have been developed to reduce their effects. In the back-end, the model applies higher probability to the statements it has seen from the training than to the ones it is generating, and there are high uncertainty errors associated with the tokens which are not present in the user-given prompt. While the standard user cannot control the effects arising from hallucinations from data and training, mitigating inference-caused errors by Faithfulness Enhanced Decoding, which emphasizes the user instruction and its alignment with the generated content [1]. The authors of [4] develop the method which prioritizes the inputs/tokens from the user context when there is a high discrepancy or uncertainty between the output of the LLM and the user's prompt. So if the given input is high enough details and does not rely on external facts, the output is less likely to have a large percentage of hallucinated content, and it is one of the reasons which emphasizes the importance of having detailed user prompts to get less error output from the ChatGPT for solving scientific problems. However, the task of generating an extremely detailed prompt is not always trivial, as there is no ending point in providing details. The amount of information and how it is structured can also influence the output of the ChatGPT. During many initial tests, even though ChatGPT-4 has a higher possibility of generating code, which works straight out of the box without regeneration, the given prompt generates complete code with ChatGPT-3.5 rather than ChatGPT-4, since it assumes that the user is asking for the structure of the code and provides only the overall framework rather than the actual implementation. A possible mitigation is either using the chain of commands or adjusting the prompt and forcing the ChatGPT to provide complete code every time, and this study employs the latter method.

A useful strategy in scientific computing is to use LaTeX-generated pseudocode as an intermediate step before code generation. This method offers experienced developers greater control by first generating a LaTeX algorithm via ChatGPT, reviewing and refining it, and then using it to produce code. This approach ensures a clear, logical structure before implementation, reducing potential errors. While LaTeX provides precision and consistency, this method introduces challenges, such as the model's need to decode LaTeX, which can complicate longer prompts or complex modifications. Additionally, this approach may not scale well for complex problems and might be less accessible for users unfamiliar with pseudocode. For proficient coders, the main advantage of ChatGPT may be time-saving rather than enhancing understanding, making this algorithmic approach not universally applicable.

6 Conclusion

The central purpose of this chapter was to explore the performance of ChatGPT-4 in generating code for complex scientific problems, which, in this case, focuses on the implementation of the 2D Fourier transient heat conduction problem using the FTCS scheme for a rectangular plate with an arbitrary number of holes. The prompt was formulated in a way such that it encapsulates every parameter used for the simulation along with the precise boundary condition, and it also highlights the use of libraries required to produce the plots requested by the user. The code generated by ChatGPT-4 was tested and verified with the analytical solution, which demonstrated that the results are in agreement with the analytical solution. Remaining errors are minor and can be traced back to the numerical method (FTCS) realized by a flawless code. The discussion elucidated the general problem of hallucination intimately related to LLMs, where we suggested effective measures to overcome these issues like providing details of the instructions in prompt. Overall, the results of this study strongly suggest that LLMs will play a major role in the future as an assistant in the development and iteration of faster solutions, thus pushing the boundaries of scientific computing and problem-solving.

References

1. Huang, L., Yu, W., Ma, W., Zhong, W., Feng, Z., Wang, H., Chen, Q., Peng, W., Feng, X., Qin, B. & Others. A survey on hallucination in large language models: Principles, taxonomy, challenges, and open questions. *ArXiv Preprint* ArXiv:2311.05232. (2023)
2. Ji, Z., Lee, N., Frieske, R., Yu, T., Su, D., Xu, Y., Ishii, E., Bang, Y., Madotto, A. & Fung, P. Survey of hallucination in natural language generation. *ACM Computing Surveys*. **55**, 1-38 (2023)
3. Oliveira, J., Nascimento, J., Ribeiro, S., Oliveira, G., Costa, G., Guimarães, G. & Santos Jr, J. Analytical solution of a 2d transient heat conduction problem using green's functions. *Revista De Engenharia Térmica*. **19**, 66-71 (2020)
4. Poel, L., Cotterell, R. & Meister, C. Mutual information alleviates hallucinations in abstractive summarization. *Proceedings Of The 2022 Conference On Empirical Methods In Natural Language Processing*. pp. 5956-5965 (2022,12), https://aclanthology.org/2022.emnlp-main.399
5. Suárez-Carreño, F. & Rosales-Romero, L. Convergency and stability of explicit and implicit schemes in the simulation of the heat equation. *Applied Sciences*. **11**, 4468 (2021)

Topics of Deep Learning Based Materials Science

Transfer Learning for Alloy Classification Based on Microstructure Images

Aditya Deshmukh and Bernhard Eidel

Abstract This study investigates the performance of large language models (LLMs), specifically GPT-4, in developing a deep learning (DL) model to predict alloy types based on scanning electron microscopy (SEM) images of steel microstructures. The approach utilizes transfer learning (TL) and an ensemble of two pre-trained models, ResNet-50 and DenseNet-121, fine-tuned on SEM scans of 33 types of steels. The two models achieve validation accuracies of 97.6% and 98.4%, respectively, with the ensemble model reaching a test accuracy of 99.2%. The results underscore the potential of LLM-assisted coding in computer vision tasks, such as image classification, within computational materials science. The limitations are also discussed.

1 Introduction

The study evaluates the ability of GPT-4 at coding a transfer-learning-based model that predicts an alloy given a Scanning Electron Microscope (SEM) scan of a microstructure. This section describes the data and approaches relevant to this task. Section 2 describes the prompt-engineering approach and the prompt. Section 3 lists the code generated by this prompt. Section 4 reports the performance of the code. Section 5 discusses the coding errors GPT-4 made and the measures taken. Finally, in Sect. 6, we summarize the findings.

A. Deshmukh (✉) · B. Eidel
M5, IMFD, TUBAF, Freiberg, Germany
e-mail: aditya.deshmukh@student.tu-freiberg.de

B. Eidel
e-mail: bernhard.eidel@imfd.tu-freiberg.de

© The Author(s), under exclusive license to Springer Nature Switzerland AG 2025
B. Eidel (ed.), *GPT for Python-Coding in Computational Materials Science and Mechanics*, Studies in Computational Intelligence 1198,
https://doi.org/10.1007/978-3-031-85470-5_6

1.1 Data

We use raw SEM scans of microstructures of 33 types of 9% Cr steels [4]. The data is divided into three folders, one for each alloy (CPJ, HR, P92), each consisting of subfolders of specific alloys (e.g., CPJ7J), wherein the microstructure images of that specific alloy are stored. These three families of alloys are closely related in terms of their functions. They are heat-resistant and exhibit good creep resistance. In total, there are 837 images and 33 alloys, with a varying number of images for each alloy. Figure 1 shows an example of the SEM scan of one microstructure.

Remark. In the context of commercial alloys, the abbreviation HR typically stands for Heat-Resistant. This designation is often used to identify alloys that are specifically designed to withstand high temperatures while maintaining their mechanical properties and corrosion resistance. The abbreviation CPJ stands for Copper-Phosphorus-Joint alloys. These are specialized brazing alloys used for joining copper and copper-based materials. The term P92 refers to a specific type of alloy used in high-temperature applications, particularly in the power generation and petrochemical industries. P92 alloy is a type of creep-resistant, ferritic-martensitic steel, known for its excellent high-temperature strength, oxidation resistance, and corrosion resistance. It is commonly used in the manufacturing of components like pipes, tubes, and fittings in high-temperature steam and pressure environments.

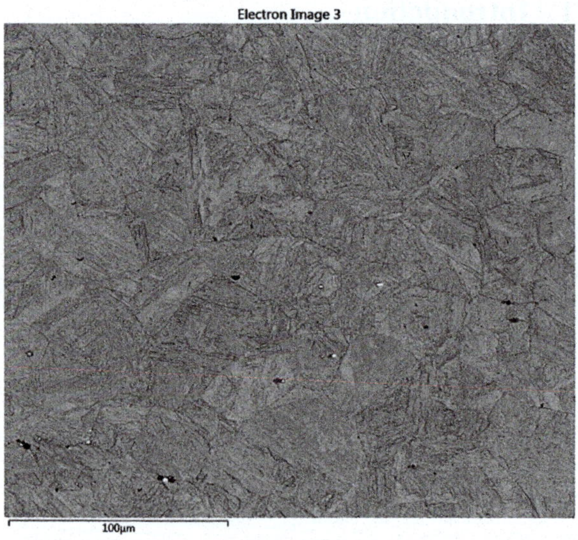

Fig. 1 Sample SEM scan of a steel microstructure. It exhibits a resolution of 630 × 592 px, while the scale bar extends to 100 μm

1.2 Deep Learning

The introduction of data science, especially deep learning (DL), to the discipline of materials science has been revolutionary [1]. DL is a subset of machine learning (ML) that uses artificial neural networks (NNs), a structure of repeating computational layers that do linear and non-linear operations to transform input data into a new useful representation and finally extract knowledge, to solve several types of problems, such as computer vision, natural language, in various disciplines, such as medicine, economics, earth sciences, and, of course, computational materials science. In computational materials science, deep learning is primarily used to accelerate the process of materials optimization and discovery, which includes problems related to forward simulation and inverse design, to better understand or exploit the property-structure-process linkages [5].

In theory, sufficiently deep DL models can approximate any function. They are therefore quite powerful. However, the goal is to not only fit the input data, called the training data. A model is considered usable only if it is generalizable, or if it also performs well on out-of-distribution data. Therefore, models are tested on test data, and the performance metrics associated with this test data indicate how well the model may perform in the wild. Training DL models also requires a third data set, called the validation set, to tune the hyperparameters associated with the model, such as the number of epochs, which is the number of times the model processes the whole dataset to learn its parameters or weights, or the learning rate, an optimization parameter that governs the step size the optimization algorithm takes. In supervised deep learning, at every optimization step, the deep learning model outputs its predictions. These are compared against the ground truth via the loss function. Minimization of loss indicates a good fit. An optimization step is taken in the direction that reduces the loss. After several epochs and consequent weight updates, the model finds the optimal weights that minimize the loss function, and the problem is considered solved.

1.3 Transfer Learning

Although the amount of data in materials science is on the rise, this increase mainly corresponds to certain subsets of materials classes and properties. Materials experiments are expensive.

Most of the big data available originates from computations; e.g., DFT databases, finite element analysis. Materials data is typically not big data for many interesting problems. This limits materials scientists from utilizing the full potential of DL.

For instance, DL cannot be used for several materials problems where the available data is scarce because DL is typically data-hungry. However, some DL models are transferable. For instance, the shallower layers of a convolutional neural network (CNN) learn more general concepts about an image. In the deeper convolutions, the kernels reflect more specific knowledge. When two problems are comparable, a DL model trained for the first problem can be used to fine-tune a model for the second problem. This is called Transfer Learning (TL), and it is an immensely useful approach when one has big data for one problem and small data for the problem at hand—a common scenario in materials science. Refer to [2] for a thorough discussion on TL.

Because the microstructures of steels look quite similar within the specific class of steel, predicting with small data alone is challenging—hence, we use TL. We use ResNet-50 and DenseNet-121, both trained on millions of images from the ImageNet data, as the two pre-trained models of the ensemble [3, 6]. ResNet-50's architecture includes residual connections that help mitigate the vanishing gradient problem, thereby improving learning in deep networks. DenseNet-121, on the other hand, features dense connectivity, where each layer is connected to every other layer, which enhances feature propagation and reduces the number of parameters, leading to improved performance. Owing to the architectural differences, the two models may produce different outputs. By combining them in an ensemble, we combine the strengths of both models.

ResNet-50 and DenseNet models are a good choice because our problem is essentially an image classification problem, and these models, having their weights trained on ImageNet, should be able to recognize the general image patterns and features. When doing TL, the pre-trained model is stacked with new layers, or, at minimum, stacked with only one layer, to match the output dimensions of the second task. Here, only the final fully connected layer is replaced by a layer that outputs a tensor of size matching the number of alloy types. Because ImageNet likely does not contain many images closely resembling alloy microstructures, we fine-tune the weights of the output layer and the inner layers. To do this, we keep the default setting for freezing, which is unfrozen. The overall model, with weights imported from the base pre-trained models and fine-tuned on our small data set, solves our problem (Fig. 2).

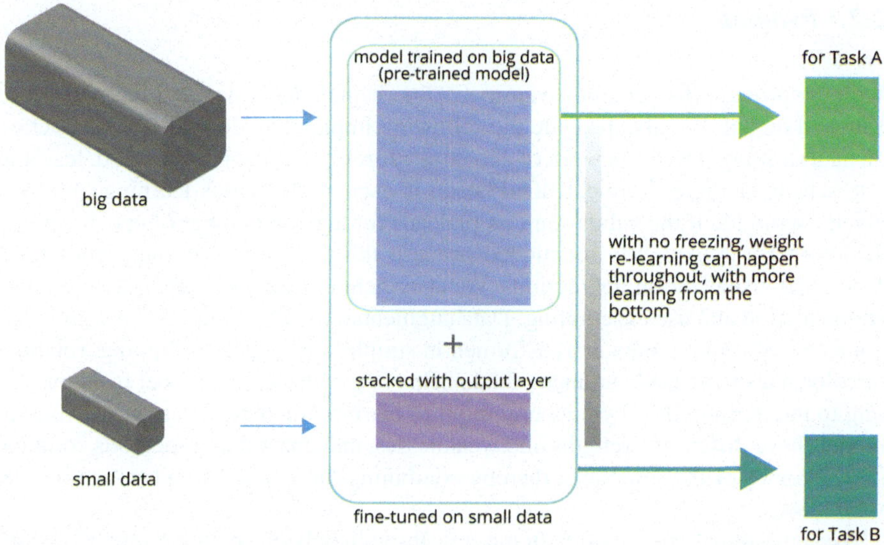

Fig. 2 Transfer Learning. A model trained for Task A on big data is used to build another model for another task, Task B, which is fine-tuned on the smaller data set. In addition to the previous model, the new model at minimum has an output layer corresponding to the output dimensions for Task B

2 Formulation of Coding Task

2.1 Prompt Engineering Approach

Before we describe the prompt, we provide a brief discussion on the prompt-engineering approach. Although a concise problem statement may be sufficient, the resulting code can have problems. At best, GPT-4 leaves the user certain programming choices and tells them when what choice is relevant and how to implement it. At worst, the user has to troubleshoot and re-prompt to fix errors. GPT-4 can be made more aware, thereby reducing such errors, by describing all user expectations, including implementation details.

If the user does not know the implementation details, such as data splitting ratio for training, validation, and test, which optimization method to use, should techniques like data augmentation be implemented, either source this information from prevalent research, or, better yet, use GPT-4. Give it the problem statement and query about the implementation details. For our problem, GPT-4 gave correct suggestions, such as data augmentation and ensemble techniques with what models to use. Include this information in the main prompt. A sufficiently detailed prompt is necessary to avoid numerous iterations of supplemental prompts and code updates, where consistency problems like changes in variable names or implementation logic may occur.

2.2 Prompt

We now provide the prompt given to GPT-4. It includes a detailed description of data and model. We also provide the following implementation details: instruction on data splitting, DL framework of choice (PyTorch), optimization method, learning rate scheduler (here, ReduceLROnPlateau is used; it decreases learning rate by a given magnitude if the validation loss plateaus for a given number of epochs), GPU device (MPS is explicitly mentioned because this program is initially built for a Mac), and visualization instructions. We also include explicit instruction to do data augmentation and use ensembling. Data augmentation is a technique wherein a random slice of training data is transformed in simple ways, such as flipping, rotation, or color. This increased variance of data helps in preventing the model from overfitting to the training data by encouraging it to learn more general features. Different models have different strengths and weaknesses, and ensembling methods combine the outcomes of these models, thereby combining the strengths and mitigating the weaknesses.

We aim to generate fully functional code through only one main prompt. However, a few supplemental prompts were necessary. These are discussed in Sect. 5.

3 Code Listing

Based on the prompt described in Fig. 3, GPT-4 generated the code listed here.

```

# Import necessary libraries
import os
import copy
import pandas as pd
import random
import numpy as np
from sklearn.model_selection import train_test_split
import torch
from torchvision import transforms, models
from torch.utils.data import DataLoader
import torch.nn as nn
import torch.optim as optim
from sklearn.metrics import accuracy_score
import matplotlib.pyplot as plt
from PIL import Image
from sklearn.preprocessing import LabelEncoder

# Set random seeds for reproducibility
random_seed = 99
torch.manual_seed(random_seed)
torch.cuda.manual_seed(random_seed)
np.random.seed(random_seed)
random.seed(random_seed)

```

```python
# Data Preparation

# Directory path
base_path = '/Users/adityadeshmukh/Desktop/RJSC'

# Alloy folders
alloy_folders = {
    'hr_alloys20220208': 'hr',
    'CPJ_alloys': 'cpj',
    'P92 OTHER': 'p92'
}

# Creating a dataframe
data = {'alloy': [], 'image': [], 'alloy type': []}

# Looping through each main folder and its subfolders
for folder, alloy in alloy_folders.items():
    folder_path = os.path.join(base_path, folder)
    for subfolder in os.listdir(folder_path):
        subfolder_path = os.path.join(folder_path, subfolder)
        if os.path.isdir(subfolder_path):
            for file in os.listdir(subfolder_path):
                if file.endswith('.bmp'):
                    file_path = os.path.join(subfolder_path, file)
                    data['alloy'].append(alloy)
                    data['image'].append(file_path)
                    data['alloy type'].append(subfolder)

# Converting to pandas DataFrame
df = pd.DataFrame(data)
print(df.head())

# Label Encoding: Convert string labels to numerical labels (which is what the model expects)
label_encoder = LabelEncoder()
df['alloy type'] = label_encoder.fit_transform(df['alloy type'])

# Splitting data
# the first line splits data into training:test sets in the ratio 70:30.
# the second line splits the test set into two halves: validation set and test set.
# so that ultimately we have train:valid:test = 70:15:15
train_df, test_df = train_test_split(df, test_size=0.3, stratify=df['alloy type'], random_state=random_seed)
val_df, test_df = train_test_split(test_df, test_size=0.5, stratify=test_df['alloy type'], random_state=random_seed)

# Data augmentation and normalization.
# Two different sets of transformations are necessary because data augmentation transformations
# only apply to the training set, not test and validation.
data_transforms = {
    'train': transforms.Compose([
```

```python
        transforms.ToTensor(),
        # Crop the top and bottom 25 pixels to remove miscellaneous SEM annotations
        transforms.Lambda(lambda x: x[:, 25:-25, :]),
        # Randomly crop and resize images to 224x224 (required for pre-trained models)
        transforms.RandomResizedCrop(224),
        # Randomly apply horizontal flipping and rotation for data augmentation:
        # Each time an image is loaded during training, the DataLoader applies the transformations randomly.
        # Essentially, this means the model sees slightly different versions of the training images throughout the
        # training process, which helps it generalize better by learning from a more diverse set of data representations.
        transforms.RandomHorizontalFlip(),
        transforms.RandomRotation(15),
        # Normalize images based on pre-defined mean and standard deviation of ImageNet dataset
        transforms.Normalize([0.485, 0.456, 0.406], [0.229, 0.224, 0.225])
    ]),
    'val': transforms.Compose([
        transforms.ToTensor(),
        transforms.Lambda(lambda x: x[:, 25:-25, :]),
        transforms.Resize((224, 224)),
        transforms.Normalize([0.485, 0.456, 0.406], [0.229, 0.224, 0.225])
    ]),
}

# Custom dataset class to load images and apply transformations
class SteelDataset(torch.utils.data.Dataset):
    def __init__(self, dataframe, transform=None):
        self.dataframe = dataframe
        self.transform = transform

    def __len__(self):
        return len(self.dataframe)

    def __getitem__(self, idx):
        img_name = self.dataframe.iloc[idx, 1]
        image = Image.open(img_name)
        label = self.dataframe.iloc[idx, 2]

        if self.transform:
            image = self.transform(image)

        return image, label


# Datasets
train_dataset = SteelDataset(train_df, transform=data_transforms['train'])
```

```
val_dataset = SteelDataset(val_df, transform=data_transforms['val'])
test_dataset = SteelDataset(test_df, transform=data_transforms['val'])

# DataLoaders in PyTorch are iterators that enable efficient loading of data during the training, validation,
# and testing phases of a machine learning model.
batch_size = 32
train_loader = DataLoader(train_dataset, batch_size=batch_size, shuffle=True, worker_init_fn=lambda _: np.random.seed(random_seed))
val_loader = DataLoader(val_dataset, batch_size=batch_size, worker_init_fn=lambda _: np.random.seed(random_seed))
test_loader = DataLoader(test_dataset, batch_size=batch_size, worker_init_fn=lambda _: np.random.seed(random_seed))

# Device configuration
# if your device has a gpu, it needs to be declared explicitly. For macbooks, the gpu is mps.
# if you use windows or linux, you should try cuda instead.
device = torch.device("mps" if torch.backends.mps.is_available() else "cpu")

# Load and modify pretrained models
def load_model(model_name, num_classes):
    if model_name == "resnet":
        # When pretrained=True, it means that the model is initialized with weights that have already been learned.
        model = models.resnet50(pretrained=True)
        # This line extracts the number of input features to the last fully connected layer (fc)
        num_ftrs = model.fc.in_features
        # This line replaces the last fc with a new fc tailored for our task by setting its output to number of classes,
        # or number of steel alloys.
        model.fc = nn.Linear(num_ftrs, num_classes)
    elif model_name == "densenet":
        model = models.densenet121(pretrained=True)
        num_ftrs = model.classifier.in_features
        model.classifier = nn.Linear(num_ftrs, num_classes)
    return model


# Number of classes
num_classes = len(df['alloy type'].unique())

# Load models
resnet = load_model("resnet", num_classes).to(device)
densenet = load_model("densenet", num_classes).to(device)
torch.manual_seed(random_seed)
torch.cuda.manual_seed(random_seed)

# Loss and optimizer
```

```
# The loss quantifies how well the model's predictions match
    ↪ labels in the training data.
# The optimizer, here stochastic gradient descent, is an
    ↪ algorithm that adjusts the parameters (weights and
    ↪ biases)
# of the neural network during training to minimize the loss.
criterion = nn.CrossEntropyLoss()
# Momentum is generally set to 0.9. LR, the learning rate,
    ↪ will be tuned.
optimizer_resnet = optim.SGD(resnet.parameters(), lr=0.001,
    ↪ momentum=0.9)
optimizer_densenet = optim.SGD(densenet.parameters(), lr
    ↪ =0.001, momentum=0.9)

# Learning rate scheduler.
# LR is tuned in this way: if the loss plateaus for 5 epochs (
    ↪ iterations), LR is reduced by a factor of 0.1
scheduler_resnet = optim.lr_scheduler.ReduceLROnPlateau(
    ↪ optimizer_resnet, 'min', factor=0.1, patience=5,
    ↪ verbose=True)
scheduler_densenet = optim.lr_scheduler.ReduceLROnPlateau(
    ↪ optimizer_densenet, 'min', factor=0.1, patience=5,
    ↪ verbose=True)

# Training function
# in machine learning, data is fed to the model as dataloaders
    ↪  for a number of iterations, called epochs.
# In each training epoch, model tries to learn optimum weights
    ↪  that minimize the training loss. This model is
    ↪ evaluated
# on validation data in a validation epoch. Hyperparameters
    ↪ are tuned in order to obtain lower validation loss.
# Accuracies are also stored for each epoch. The model
    ↪ associated with the best validation
# accuracy is returned, along with the entire history of
    ↪ training loss, training accuracy, validation loss, and
# validation accuracy.
def train_model(model, dataloaders, criterion, optimizer,
    ↪ scheduler, num_epochs=25):
    best_acc = 0.0
    # Initialize arrays to store metrics
    train_losses, val_losses, train_accs, val_accs = [], [],
    ↪ [], []

    for epoch in range(num_epochs):
        print('Epoch {}/{}'.format(epoch, num_epochs - 1))
        print('-' * 10)

        # Each epoch has a training and validation phase
        for phase in ['train', 'val']:
            if phase == 'train':
                model.train()  # Set model to training mode.
    ↪ Here weights are learned.
            else:
                model.eval()   # Set model to evaluate mode.
    ↪ Here we only use the learned weights to make
    ↪ predictions.
```

```
            running_loss = 0.0
            running_corrects = 0

            # Iterate over data
            for inputs, labels in dataloaders[phase]:
                inputs = inputs.to(device)
                labels = labels.to(device)

                # Zero the parameter gradients
                optimizer.zero_grad()

                # Forward
                with torch.set_grad_enabled(phase == 'train'):
                    outputs = model(inputs)
                    _, preds = torch.max(outputs, 1)
                    loss = criterion(outputs, labels)

                    # Backward + optimize only if in training
↪ phase
                    if phase == 'train':
                        loss.backward()
                        optimizer.step()

                # Statistics
                running_loss += loss.item() * inputs.size(0)
                running_corrects += torch.sum(preds == labels.
↪ data)

            epoch_loss = running_loss / len(dataloaders[phase
↪ ].dataset)
            epoch_acc = running_corrects.float() / len(
↪ dataloaders[phase].dataset)

            # Record metrics
            if phase == 'train':
                train_losses.append(epoch_loss)
                train_accs.append(epoch_acc.item())
            else:
                val_losses.append(epoch_loss)
                val_accs.append(epoch_acc.item())
                scheduler.step(epoch_loss)  # Adjust learning
↪ rate based on validation loss

            print('{} Loss: {:.4f} Acc: {:.4f}'.format(phase,
↪ epoch_loss, epoch_acc))

            # Deep copy the best-performing model
            if phase == 'val' and epoch_acc > best_acc:
                best_acc = epoch_acc
                best_model_wts = copy.deepcopy(model.
↪ state_dict())

        print()

    print('Best val Acc: {:4f}'.format(best_acc))
```

```
    # Load best model weights
    model.load_state_dict(best_model_wts)

    return model, train_losses, val_losses, train_accs, val_accs

# this function plots train_losses, val_losses and train_accs, val_accs in two graphs.
def plot_performance(train_losses, val_losses, train_accs, val_accs, file_name):
    np.random.seed(random_seed)
    plt.figure(figsize=(12, 5))

    plt.subplot(1, 2, 1)
    plt.plot(train_losses, label='Training loss')
    plt.plot(val_losses, label='Validation loss')
    plt.title('Training and Validation Loss')
    plt.xlabel('Epochs')
    plt.ylabel('Loss')
    plt.legend()

    plt.subplot(1, 2, 2)
    plt.plot(train_accs, label='Training accuracy')
    plt.plot(val_accs, label='Validation accuracy')
    plt.title('Training and Validation Accuracy')
    plt.xlabel('Epochs')
    plt.ylabel('Accuracy')
    plt.legend()
    plt.savefig(file_name)


# Train models
dataloaders = {'train': train_loader, 'val': val_loader}
print('\n RESNET-50________________________________ \n')
resnet, resnet_train_loss, resnet_val_loss, resnet_train_acc, resnet_val_acc  = train_model(resnet, dataloaders, criterion, optimizer_resnet, scheduler_resnet, num_epochs=500)
print('\n DENSENET-121_____________________________ \n')
densenet, densenet_train_loss, densenet_val_loss, densenet_train_acc, densenet_val_acc = train_model(densenet, dataloaders, criterion, optimizer_densenet, scheduler_densenet, num_epochs=500)

torch.save(resnet.state_dict(), 'resnet_model.pth')
torch.save(densenet.state_dict(), 'densenet_model.pth')

# Ensemble
# we average the outputs of our two models and use this averaged output to make predictions.
class AveragingEnsemble(nn.Module):
    def __init__(self, modelA, modelB):
        super(AveragingEnsemble, self).__init__()
        self.modelA = modelA
        self.modelB = modelB

    def forward(self, x):
```

```
        # Get predictions from both models
        outputA = self.modelA(x)
        outputB = self.modelB(x)

        # Average the predictions
        average_output = (outputA + outputB) / 2
        return average_output

ensemble_model = AveragingEnsemble(resnet, densenet)


# Evaluate Ensemble
def evaluate_model(model, dataloader):
    model.eval()
    all_preds = []
    all_labels = []
    with torch.no_grad():
        for inputs, labels in dataloader:
            inputs = inputs.to(device)
            labels = labels.to(device)
            outputs = model(inputs)
            _, preds = torch.max(outputs, 1)
            all_preds.extend(preds.tolist())
            all_labels.extend(labels.tolist())

    accuracy = accuracy_score(all_labels, all_preds)
    return accuracy


test_accuracy = evaluate_model(ensemble_model, test_loader)
print('Test Accuracy of Ensemble: {:.4f}%'.format(
    ↪ test_accuracy * 100))

# Plotting functions
plot_performance(resnet_train_loss, resnet_val_loss,
    ↪ resnet_train_acc, resnet_val_acc, "res1.png")
plot_performance(densenet_train_loss, densenet_val_loss,
    ↪ densenet_train_acc, densenet_val_acc, "res2.png")
```

Code Listing 6.1 Program by ChatGPT-4 for classifying microstructure

I require a Python program for classifying steel microstructure images according to their alloy type using transfer learning. A description of the data as well as the model is provided below.

Data

I have three folders of steel microstructure images in the directory /Users/adityadeshmukh/Desktop/RJSC. Each folder is described here as follows:

- The folder 'hr_alloys20220208' contains the initial microstructure for HR alloys.
  - It contains 11 subfolders with subfolder titles corresponding to a specific HR alloy.
  - The '.bmp' files in each subfolder are the raw microstructure images for that specific alloy.
- The folder 'CPJ_alloys' contains the initial microstructure for CPJ alloys.
  - It contains 18 subfolders with subfolder titles corresponding to a specific CPJ alloy.
  - The '.bmp' files in each subfolder are the raw microstructure images for that specific alloy.
- The folder 'P92 OTHER' contains the initial microstructure for P92 alloys.
  - It contains 4 subfolders with subfolder titles corresponding to a specific P92 alloy.
  - The '.bmp' files in each subfolder are the raw microstructure images for that specific alloy.

Note that all three folders and subfolders have differently named and different numbers of items. You'd first make a pandas dataframe with three columns: 'alloy' (cpj, hr, or p92), 'image' (path to the .bmp image file), 'alloy type' (name of the subfolder wherein the image exists, which is also the specific alloy).

Model

- Transfer learning requires a pre-trained model. Comment on the feasibility of ResNet-50 and DenseNet-121 for this task. Fine-tune on all two and create an ensemble.
- Use PyTorch framework.
- Set GPU device as 'mps'.
- While fine-tuning, ensure that there is no freezing.

Fig. 3 Prompt for classifying microstructures

- Split the data into training (70%), validation (15%), and test (15%) sets stratified according to the label 'alloy type.' Use train_test_split twice to achieve this.
- Do data augmentation.
- Use the validation loss to find the optimum number of epochs. Also, use ReduceLROnPlateau scheduler to tune the learning rate. Set the number of epochs to 500 for each model.
- Record training loss, training accuracy, validation loss, validation accuracy for each epoch for each model. Ultimately, record test accuracy for the ensemble. Write code to make relevant graphs.

Do not give me only the code skeleton; I want the whole program in one code block.

Fig. 3 (continued)

4 Performance

Figures 4 and 5 show the performance of the fine-tuned ResNet-50 model and the fine-tuned DenseNet-121 model respectively. For both models, with an increasing number of epochs, the training loss and validation loss decrease, and consequently the training accuracy and validation accuracy increase. At the last epoch, we observe a validation accuracy of 97.6190% and 98.4127% for ResNet-50 and DenseNet-121 respectively, as shown in Table 1, with the test accuracy of the ensemble being 99.2063%.

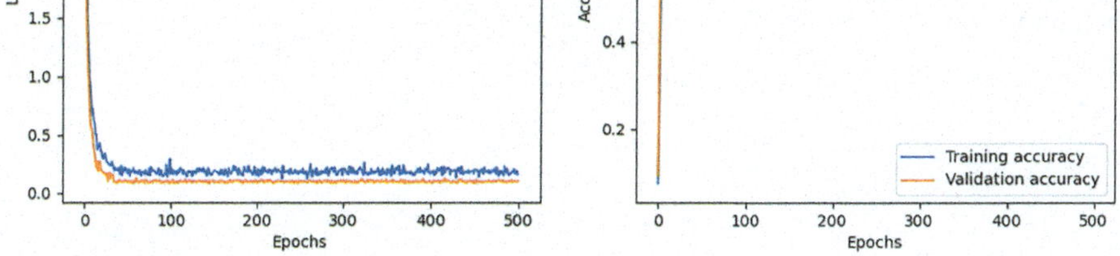

Fig. 4 Performance of ResNet50 Model. (Left:) Training and validation loss. (Right:) Training and validation accuracy

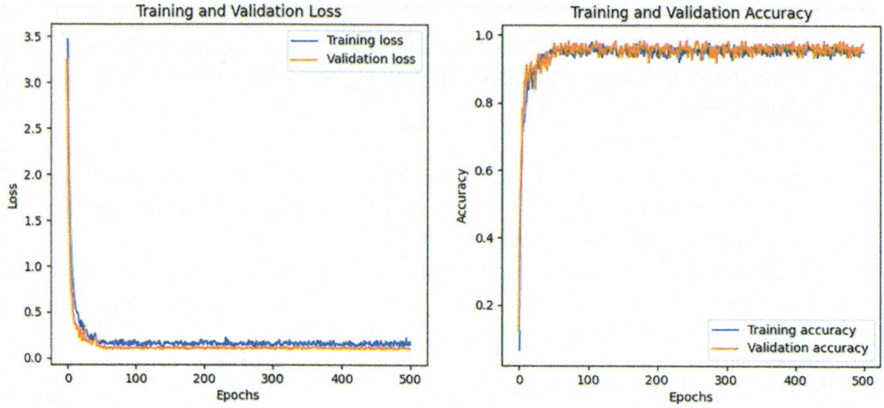

Fig. 5 Performance of DenseNet121 Model. (Left:) Training and validation loss. (Right:) Training and validation accuracy

Table 1 Model performance comparison

	ResNet-50	DenseNet-121	Ensemble
Validation Acc	97.6190%	98.4127%	
Test Acc			99.2063%

Figure 6 displays a random sampling of 12 images from the test set. The ground truths and the predicted labels are in good agreement. Note that the images appear differently from the sample image shown in Fig. 1 because of ImageNet normalization; this is a necessary step in pre-processing when using pre-trained models like ResNet-50.

The observation that the validation loss is almost consistently smaller than the training loss in the left of Figs. 4 and 5 can have different reasons. Augmented data may have been harder for the model to learn and because data augmentation transformations happen only on some samples belonging to the training set and not at all on the validation set, the validation results remain unaffected. Smaller datasets (validation) have smaller intrinsic variance than larger datasets (training); recall their ratio of 70:15. Or this is simply an accident in that the data split is suitable for such behavior.

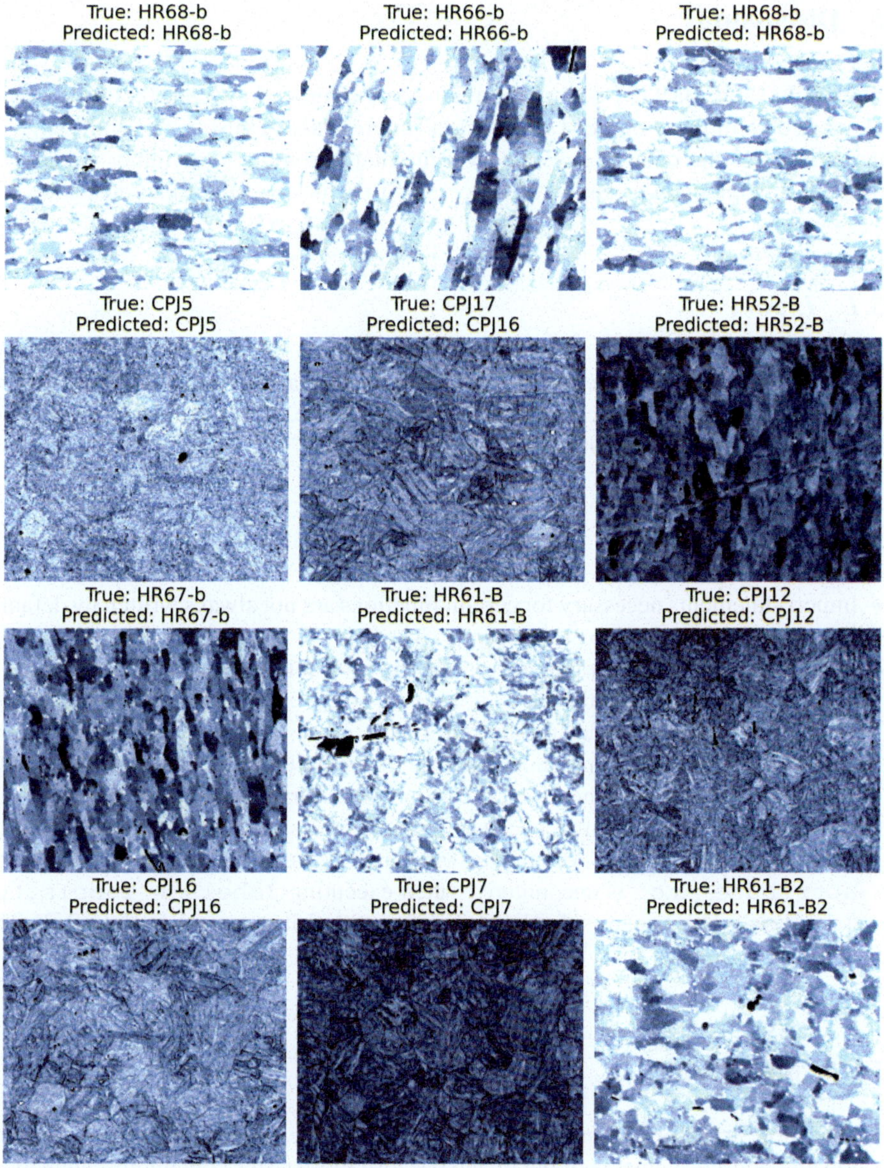

Fig. 6 A random sampling of 12 images from the test set with their ground truths and predicted labels

5 Discussion

The results show that GPT-4 is successful in generating working code for the problem of image classification on SEM scans of steel microstructures using TL. While almost the entirety of the code is generated via the main prompt, some minor tweaking, through supplemental prompts, was necessary to achieve a fully functional code.

5.1 *Errors*

The errors are described below in three categories: major errors, i.e. errors which require a correction for the code to run; minor errors, i.e. errors with which code will run but perform sub-optimally; and ignored tasks.

5.1.1 Major Errors

- Import statements necessary for certain functions are not always written by default (e.g., `from PIL import Image`).
- While creating the dataframe, traversing the relevant directories requires exception handling to ignore the `Not a directory` error due to the presence of `.DS_Store` files that are present on a Mac. Note that prior handling of such an error is a foresight that should not be expected of LLMs because this is an exception, not a norm. Therefore, this is not necessarily a drawback.
- Label encoding is required because original labels are strings when PyTorch requires numbers. `AttributeError: 'tuple' object has no attribute 'to'` is encountered when executing `labels = labels.to (device)` This suggests that labels is a tuple when it is expected to be a PyTorch tensor.

5.1.2 Minor Errors

- One more transformation is necessary in image pre-processing to crop out the top and bottom segments of the image because those strips contain miscellaneous SEM annotations. GPT-4 cannot–and is not expected to–fix such specific errors it is not even aware of. However, the fix, described later, is straightforward.
- Learning rate scheduler step is initially taken on training loss instead of validation loss.
- The initial ensemble model uses a neural network with a linear layer that combines the individual outputs of the two models, ResNet-50 and DenseNet-121, and ReLU. It performs quite poorly. It is replaced with an averaging operation upon request.

- Default values for the parameters of the LR scheduler are used. We test several models to tune these values. GPT-4 does not provide unique instruction to guide these values, but optimization requires scanning the hyperparameter space with repeated tries. This obviously falls beyond what LLMs can do on their own.

5.1.3 Ignored Tasks

Training loss, training accuracy, validation loss, and validation accuracy for each epoch is not recorded. The visualization task is fully ignored.

5.2 *Fixes*

The above errors have easy fixes. GPT-4 is good at troubleshooting; one only needs to copy paste the error encountered by the Python IDE into GPT-4. GPT-4 will either give one or more potential solutions. Some errors are quite elusive. GPT-4 does not right away figure out the case for label encoding. It first examines SteelDataset class's `__getitem__` method and unpacking of labels inside training loop. When we explicitly state that both of these are correct, it lists more possibilities. One of these is label transformation, which reminds us that the labels are originally in strings, when PyTorch requires labels to be numerically encoded. GPT-4 finally suggests to use label encoding. This is a process; however, it could have been avoided by pointing this out right in the main prompt. When fixing other problems, such as replacing ensemble function or obtaining code for previously ignored task, a simple instruction is usually sufficient. Table 2 lists these supplemental prompts.

In addition to these fixes, the code is slightly modified to make results reproducible; as long as the random seed is left unchanged, whose purpose is to maintain consistency in random numbers generated in the program, one gets the same results every time. Please note that we train the model for 500 epochs to observe if any learning happens over the long run. However, similar results are achievable with only 50 epochs.

Table 2 Modifications and supplemental prompts

Modification	Supplemental prompt
New ensemble function	Give me an ensemble that simply gives the average
LR step correction	Fix this to take scheduler step based on validation loss
Visualization function	I need to store arrays of validation loss, validation accuracy, training loss, training accuracy for each epoch so that I can print them later. Also implement that and give me code to visualize the performance

6 Conclusion

We obtain a working code through GPT-4 to predict an alloy from an SEM scan of its microstructure. The resulting model is an ensemble of two pre-trained models, ResNet-50 and DenseNet-121, fine-tuned on a training set that includes SEM scans of microstructures of 33 different types of 9% Cr steels. The model achieves an accuracy of 99.20% on the test set, indicating that it performs quite well.

GPT-4 is good at unpacking a prompt, understanding it, writing code, explaining the code in a broad sense as well as line-by-line, when asked to, and the generated code is also well documented. It helps to have a sufficiently detailed prompt, rich with information on not only the nature of the problem but also implementation details. Troubleshooting is easy by simply providing GPT-4 with the traceback of the errors encountered. Code that needs to be modified or rewritten can be edited with additional simple prompts.

By solving this problem, we demonstrate that, with human intervention, GPT-4 is capable of generating code for common computer vision tasks such as image classification in computational materials science.

References

1. Agrawal, A. & Choudhary, A. Perspective: Materials informatics and big data: Realization of the "fourth paradigm" of science in materials science. *Apl Materials.* **4** (2016)
2. Zhuang, F., Qi, Z., Duan, K., Xi, D., Zhu, Y., Zhu, H., Xiong, H. & He, Q. A Comprehensive Survey on Transfer Learning. *Proceedings Of The IEEE.* **109**, 43-76 (2021)
3. He, K., Zhang, X., Ren, S. & Sun, J. Deep residual learning for image recognition. *Proceedings Of The IEEE Conference On Computer Vision And Pattern Recognition*. pp. 770-778 (2016)
4. Rozman, K., Doğan, Ö., Chinn, R., Jablonksi, P., Detrois, M. & Gao, M. Dataset for machine learning of microstructures for 9% Cr steels. *Data In Brief.* **45** pp. 108714 (2022)
5. Choudhary, Kamal, et al. Recent Advances and Applications of Deep Learning Methods in Materials Science. *Npj Computational Materials*, vol. 8, no. 1, 2022. pp. 1-26. www.nature.com, https://doi.org/10.1038/s41524-022-00734-6.
6. Huang, G., Liu, Z., Van Der Maaten, L. & Weinberger, K. Densely Connected Convolutional Networks. *2017 IEEE Conference On Computer Vision And Pattern Recognition (CVPR)* https://doi.org/10.1109/cvpr.2017.243

Transfer Learning for Microstructure Image Segmentation

Rahul Narkhede and Bernhard Eidel

Abstract Microstructure segmentation is a crucial task in materials science which facilitates detailed material characterization and establishing processing-structure-property linkages. Image segmentation is a classical technique in computer vision for which several deep learning models exist. Furthermore, transfer learning can leverage the performance of these deep learning models on new datasets by initializing them with parameters pre-trained on large datasets. In this chapter, transfer learning is applied for segmentation of nickel-based superalloy microstructure images using a model pre-trained on a large dataset of microscopy images called MicroNet. GPT-4 on ChatGPT Plus is instructed to generate a Python code for performing this task. By a suitably designed sequence of prompts, GPT-4 provides promising results in implementing the task. Aspects of prompt design, handling errors and testing the outcome are considered as well.

1 Introduction

Image segmentation is a cornerstone technique in visual data analysis, enabling the differentiation of objects within an image based on distinct characteristics. This technique is particularly crucial in materials science for the quantification and analysis of microstructures, which is fundamental for understanding the properties and behaviors of materials. Traditionally, segmentation has been performed manually or through semi-automated methods, which can be labor-intensive and prone to error. With the advent of computer vision and machine learning, more sophisticated and auto-

R. Narkhede · B. Eidel (✉)
M5, IMFD, TUBAF, Freiberg, Germany
e-mail: bernhard.eidel@imfd.tu-freiberg.de

R. Narkhede
e-mail: rahul-vishnu.narkhede@student.tu-freiberg.de

© The Author(s), under exclusive license to Springer Nature Switzerland AG 2025
B. Eidel (ed.), *GPT for Python-Coding in Computational Materials Science and Mechanics*, Studies in Computational Intelligence 1198,
https://doi.org/10.1007/978-3-031-85470-5_7

mated methods have been developed, significantly enhancing accuracy and efficiency [1, 2].

Transfer learning is a powerful strategy in machine learning where a model developed for one task is repurposed for a second, related task. This approach is especially valuable in environments where data are scarce or where training a new model from scratch is computationally prohibitive. By utilizing models pre-trained on large datasets, transfer learning not only conserves resources but also enhances model accuracy, making it a vital tool in domains where data acquisition is challenging [3].

Among the several deep learning architectures available, the ResNet50 [4] architecture is known for its depth and the use of residual connections, which enabled training significantly deeper networks than previously feasible. This architecture is adept at extracting detailed features that are crucial for accurate segmentation, and is hence often employed as an encoder in segmentation tasks. The encoder transforms input images to rich, feature-dense representations for the model to capture the image context. Then, these representations are translated back into meaningful segmentation maps using decoder architectures like the UNet++. UNet++ [5] is an improvement of the classic UNet architecture, and shows improvement in segmentation accuracy. The overall framework of the microstructure segmentation task is shown in the Fig. 1.

The task in this chapter is to apply transfer learning for segmentation of microstructure images of nickel-based superalloys into matrix, secondary and tertiary phases. In doing so, it uses the ResNet50 architecture as the encoder and the UNet++ as the decoder. The encoder is pre-trained on a large microscopy dataset named MicroNet. The application of a pre-trained encoder thus implies use of transfer learning for the segmentation of nickel-based superalloy dataset. The work in this chapter is largely based on the example notebook for image named `multiclass_segmentation_example.ipynb` provided as supplementary material with [6]. This segmentation task encompasses the following key steps:

- **Data Preparation**: Automated extraction and association of images with corresponding masks to establish ground truths for supervised learning.
- **Data Augmentation**: Application of transformations such as rotations and flips, and adjustments in brightness and contrast to enhance model robustness.
- **Model Setup**: Building a model based on the ResNet50 encoder pre-trained on the MicroNet dataset and a UNet++ decoder.
- **Loss function**: Integration of Dice Loss and Binary Cross Entropy (BCE) Loss to optimize both the segmentation overlap and pixel-wise classification accuracy.
- **Performance Metric**: Adoption of Intersection over Union (IoU) to quantify model performance, a standard metric in segmentation model evaluation
- **Model Training and Early Stopping**: Implementation of early stopping based on IoU improvement to prevent overfitting.
- **Testing and Visualization**: Evaluation of the model on unseen data and visualization of the results to assess performance qualitatively.

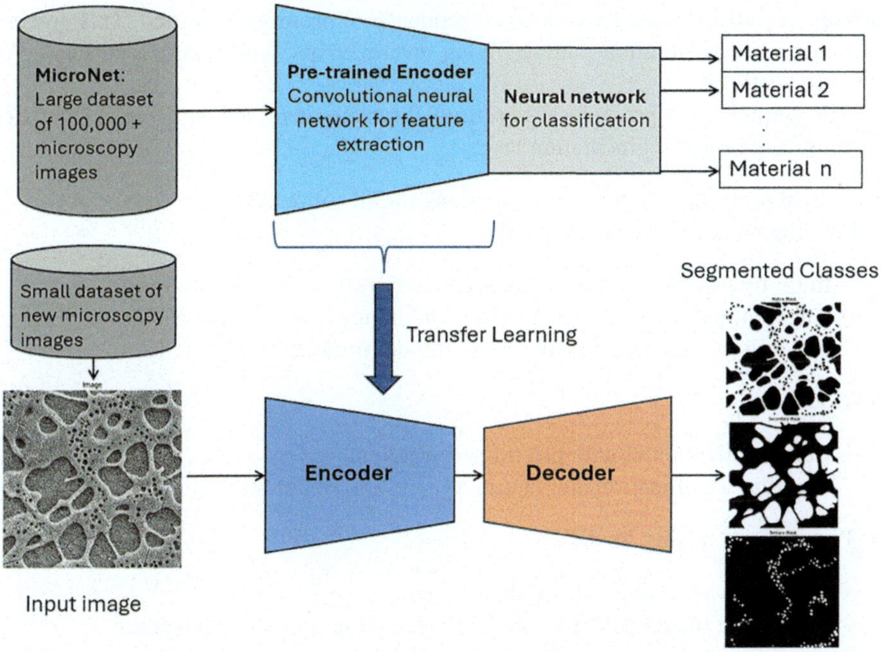

Fig. 1 Schematic showing the use of encoder pre-trained on the MicroNet dataset for microstructure image segmentation via transfer learning. The encoder (top), like ResNet50, is a convolutional neural network followed by a dense neural network classifier (shown in the gray rectangular box on the top), which is trained to classify microscopy images into different material classes. The encoder is pre-trained on the MicroNet dataset containing more than 100,000 microstructure images. By transfer learning, the pre-trained encoder is then used in the encoder-decoder segmentation model (shown in the bottom), which is then trained on a small dataset of microstructure images to segment the microstructure into separate classes

2 Prompt

The Python code for the segmentation of microstructure images is generated by GPT-4. This task requires pre-processing of data, building the deep learning model with pre-trained weights, implementation of a training routine with multiple intricacies, evaluating model performance and visualizing the predictions. It also involves implementing methods from multiple Python libraries. Considering these aspects, it is decided to instruct GPT-4 in multiple sub-tasks which include the details of all the mentioned steps. Moreover, it is also observed that upon providing all the details of this segmentation task as a one-thrust prompt resulted in generation of incomplete Python which also overlooked some instructions. Using fundamental aspects of prompt engineering, and at times using GPT-4 to summarize its steps in a refined prompt, a well-instructed and detailed sequence of prompts is drafted. This process involved multiple iterations and resolving several errors, which are also discussed in the next sections.

The sequence of prompts is based on the steps involved in a typical transfer learning and image segmentation task:

- System prompt with general instructions for all sub-tasks
- Loading and pre-processing data
  - Importing microstructure images and assigning classes to the masks.
  - Augmenting the dataset by randomized image transformations.
  - Loading the augmented data for the model training and inference.
- Model setup
  - Building the model with pre-trained weights.
  - Creating loss function and metric to measure model performance.
- Training and testing model
  - Setting up a training and validation loop.
  - Evaluating model performance and visualizing prediction accuracy.

Considering these steps, six prompts have been used to generate the desired Python code for the segmentation task:

- **Prompt 1**: System prompt
- **Prompt 2**: Dataset configuration
- **Prompt 3**: Data augmentation and data loading
- **Prompt 4**: Model setup, loss function and metric
- **Prompt 5**: Training loop setup
- **Prompt 6**: Model testing and visualizing predictions.

2.1 System Prompt

The system prompt orients the chatbot to the task. In this system prompt, ChatGPT 4 is instructed to adapt a persona which knows the requisite libraries and knowledge to execute the segmentation task. Through such a system prompt, the chatbot is also given specific instructions to mitigate persistently observed errors in the generated code.

In response to prompt shown in Fig. 2, ChatGPT 4 acknowledges that it is prepared to help the user with the task and that it anticipates further instructions.

Prompt 1
You are an expert computer vision and machine learning engineer. You have the fundamental knowledge of deep learning, the methods and their working, popular deep learning architectures for image classification and segmentation. You are specifically well versed in PyTorch and allied libraries like albumenations, segmentation_models_pytorch, DataLoader and so on. You can load pre-trained models from urls and fine-tune them for the specific task at hand. You write an error-free code. Specifically, your code encounters no issues in the dimensions of the various tensors and arrays in the deep learning process. You analyze code for any errors before presenting it. You generate complete code with documentation. You write code for the part that you are asked for, no further steps. Only what's specifically instructed. Do not show sample usage until asked for. You track matrix dimensions throughout the code and handle any dimension errors. You also ensure that any torch tensor is being moved to the CPU before it is converted to a NumPy array. Any array dimension errors will leave a poor score on your performance. The task I want to achieve is complex and I shall tell you the instructions in a step-wise fashion.

Fig. 2 System prompt provided to ChatGPT 4 before specific instructions for the segmentation task

2.2 Loading and Pre-processing Data

As the first and one of the crucial steps of the segmentation task, ChatGPT 4 is provided all the details to load the data in a suitable format for further processing and applying deep learning model for segmentation.

The prompt begins with a brief description of the overall segmentation task with details about the model architecture. This short description acts as a context for the chatbot in its subsequent responses. It is also asked to add randomization seeds to ensure that the results are reproducible.

The location of the training, validation and testing data is specified in the prompt. The dimensions of the images and masks and their file formats are also specified. Specific code instructions to read the images and masks, and then to assign classes to the masks are given. The detailed instructions are important to maintain the correct assignment of annotated masks to the images.

Pre-processing the images involves data augmentation and normalizing the image data. Data augmentation in image segmentation involves applying transformations like flipping, rotation, and adjustments to contrast and brightness to existing images, enhancing dataset diversity and model generalization without increasing the number of original images. This method helps prevent overfitting, especially with smaller or less diverse datasets. After augmentation, the images are normalized using the mean and standard deviation of the pre-trained model, in this case, the 'ResNet50' model pre-trained with the 'imagenet' dataset.

Prompt 2
Dataset Configuration:

- Paths to dataset:
  - Base directory: `'./Super1'`.
  - Training images: `'base_directory/train'`
  - Training masks: `'base_directory/train_annot'`
  - Validation images: `'base_directory/val'`
  - Validation masks: `'base_directory/val_annot'`
  - Test images: `'base_directory/test'`
  - Test masks: `'base_directory/text_annot'`
- Image and mask formats: Both image and masks are in '.tif' format with the shape (512, 512,3). Please note that the image name is for e.g. 'spot3_0d_3.tif' and its corresponding mask name is 'spot3_0d_3_mask.tif'.
- Include a custom dataset class for handling image and mask loading, applying transformations and parsing class-specific masks.
- Use `cv2.cvtColor` with `cv2.COLOR_BGR2RBG` argument for reading images. Use `cv2.imread(mask_path, 1)` to read the masks. Do not use `cv2.COLOR_BGR2RGB` as it changes their sequence.
- Extract the classes from the masks. Use the following mask colors and assign them to the classes in the dataset. Mask class colors:
  - Matrix: `'[0, 0, 0]'`
  - Secondary: `'[255, 0, 0]'`
  - Tertiary: `'[0, 0, 255]'`

  Create a binary mask for each class and assign the names 'matrix', 'secondary' and 'tertiary' by checking the class colors provided above. Then update the first mask again, i.e. 'matrix', since it is formed by all the remaining pixels after extracting the 'secondary' and the 'tertiary' classes.
- Use this stacked mask array for the classes of the masks in the rest of the code.

Fig. 3 Prompt with instructions to get the data

The pre-processed data is then passed to a dataloader. Dataloaders efficiently manage the flow of data during training by batching, shuffling, and preparing data for input into the model. They handle loading data from the dataset into memory, applying transformations such as data augmentation on-the-fly, and ensuring that data is supplied to the model in suitable format for efficient processing, thus optimizing the use of computational resources.

The specific instructions for pre-processing the data and then creating dataloaders for training, validation and testing are provided to ChatGPT 4 in the next prompt. The detailed transformations for pre-processing the data Fig. 4 and the instructions to load the final pre-processed data for model training, testing and validation Fig. 5 are provided in prompt 3.

Prompt 3
Data Augmentation: Apply transformations using the 'albumenations' library as per the instructions below:

- Create a function `'get_training_augmentation‘` which augments the data using a series of image transformations and then normalizes the data. Here is a breakdown of the transformations included in the pipeline:
  - Flipping: Images are flipped horizontally with a probability of 0.75.
  - Rotation: Images are randomly rotated by 90 degrees with a probability of 1, meaning every image is rotated.
  - Gauss Noise: Gauss noise is added to images with a probability of 0.5 to mimic image artifacts and sensor noise.
  - Contrast and brightness adjustments:
    - CLAHE: Enhances the contrast of the images.
    - Random Brightness and Contrast: Adjusts the brightness with a limit of 0.25 and modifies contrast, applied randomly.
    - Random Gamma: Adjusts the gamma values of the image.

    These adjustments are grouped in a OneOf block, meaning that one of these transformations will be randomly applied with a probability of 0.50.
  - Sharpness and blurring:
    - Sharpen: Enhances the sharpness of the images.
    - Blur: Applies a blurring effect with a blur limit of 3.

    These are also in a OneOf block with a probability of 0.50, ensuring that one of these transformations is applied to each selected image.
  - Color adjustments:
    - Random Brightness and Contrast: Further adjusts brightness and contrast with a specific limit on contrast.
    - Hue and Saturation: Modifies the hue and saturation values of the images.

    This set is another OneOf arrangement with a probability of 0.50.
  - Normalization: Standardizes the pixel values across the dataset using mean values and standard deviations of 'imagenet' dataset for the 'resnet50' encoder. Use functions from `segmentation_models_pytorch` to get the mean and standard deviation values.
  - Tensor Conversion: Converts the images from NumPy arrays to PyTorch tensors, facilitating their usage in deep learning models.

    The entire set of transformations is composed into a pipeline using

Fig. 4 Part one of prompt 3 which provides instructions for pre-processing the data

> `A.Compose`, which sequentially applies these transformations to the training images.
> The validation and testing dataset do not go through this augmentation. These two datasets only need to be normalized using the same parameters as the training dataset. Create a function named `'get_validation_augmentation'`, which only includes this normalization step and the conversion to the PyTorch tensors.

Fig. 4 (continued)

> **Prompt 3 (continued)**
> **Data Loading:**
>
> - Use the custom class for handling data and the `'get_training_augmentation'` for the dataset augmentation, to create train, test and validation datasets.
> - Use the `DataLoader` library to create a train loader with a batch size of 4, a validation loader with a batch size of 4 and a test loader with a batch size of 1. Use multiple `num_workers` to reduce memory requirement during training.
> - These loaders will be used to load data during training the model.
>
> **Visualization:** Create a function that visualizes any four of the augmented images and their masks. Next to each sample image, the three masks: 'matrix', 'secondary' and 'tertiary' should be plotted. Ensure that the image and mask dimensions are checked before the plotting and transformed suitably.

Fig. 5 Part two of prompt 3 which provides instructions for loading the data and visualizing some images with their masks

2.3 Model Setup

For the microstructure image segmentation, the 'resnet50' encoder is used with the 'UNet++' segmentation model. The pre-trained encoder weights are downloaded from a url in the model setup. The function `get_pretrained_microscopy_url` that generates the url is used as is from the source [6]. In the prompt 4 (Fig. 6), these details about the architecture and the source url to obtain the pre-trained encoder weights are provided, along with further details about the inputs for the model setup, the activation function of model output, usage of GPU if available and the expected output.

To track the loss over training the model, a custom loss function as used in the example notebook `multiclass_segmentation_example.ipynb` that combines the Dice and binary cross-entropy (BCE) losses is then created. The model performance is checked using the intersection over union (IoU) metric, which can be obtained from the '`segmentation_models_pytorch`' library. However, it is noticed that ChatGPT 4 is not aware of the latest version of the IOU function from this library. To overcome the persistent error of incorrect syntax in using the IOU metric, an example usage based on the syntax of the latest version is provided in the loss functions and metrics part of the prompt 4 shown in Fig. 7.

Prompt 4
Model Setup

- Objective: Write a Python function named `setup_segmentation_model` that initializes a UNet++ segmentation model with a custom pre-trained ResNet50 encoder. The function should handle model downloading, initialization, and setting device compatibility for running either on CPU or GPU.
- Inputs:
  - encoder_name: A string indicating the type of encoder, default to 'resnet50'.
  - class_values: A list containing unique class identifiers for segmentation tasks.
  - encoder_weights: 'micronet' or 'image-micronet'. These are the two dataset names on which the pre-trained weights can be made available.
- Model Initialization:
  - Initialize a UNet++ model with the provided encoder_name.
  - Set encoder_weights to None to skip default weight loading.
  - The model should take inputs with 3 channels and output the number of classes based on the length of class_values.
  - Use 'softmax2d' for the output layer activation if multiclass segmentation. For binary segmentation use 'sigmoid' function.
- Device Compatibility: Determine if CUDA is available for GPU utilization; if not, default to CPU. Accordingly set the map for loading the model.
- Load Custom Weights:
  - Generate url to load model using the following function: `get_pretrained_microscopy_url('resnet50', encoder_weights)`. I have the function already with me. No need to code it. Just use it.
  - Then set the encoder of the UNet++ architecture with the pre-trained encoder weights by using: `model.encoder.load_state_dict(model_zoo.load_url(url, map_location=map))`
- Output: Return the initialized and configured model and the device on which the model is set up, ready for training or inference.

Fig. 6 Prompt to set up the model

Prompt 4 (Continued)
Loss Function and Metrics

- Implement a combined Dice and Binary Cross-Entropy (BCE) loss function with a weighing factor towards BCE loss. The inputs must be processed through a sigmoid function and the BCE loss should be computed using torch.nn.functional library's `binary_cross_entropy_with_logits` function. To reshape the inputs and targets, use reshape method.
- Define and implement an Intersection over Union (IoU) metric for performance evaluation. This metric should be applicable to either case of binary or multiclass classification. You can use the `'segmentation_models_pytorch'` library for this. An example of their IoU metric function is as follows.

```
import segmentation_models_pytorch as smp

# lets assume we have multilabel prediction for 3 classes
output = torch.rand([10, 3, 256, 256])
target = torch.rand([10, 3, 256, 256]).round().long()%

# first compute statistics for true positives, false  positives, false negative and
# true negative "pixels"
tp, fp, fn, tn = smp.metrics.get_stats(output, target, mode='multilabel', threshold=0.5)

# then compute metrics with required reduction (see metric docs)
iou_score = smp.metrics.iou_score(tp, fp, fn, tn, reduction="micro")

```

- Use the loss function and the IoU function in the training of the model as I will instruct you in the next steps.

Fig. 7 Prompt to create the loss function and the model evaluation metric

2.4 Training the Model

After setting up the training, validation and test data, the model, the loss function and the metric, a function to train the model is required. This function shall put the individual components together, and create the training loop. As shown in the prompt in Fig. 8, ChatGPT 4 is instructed to set up an optimizer, use the losses as the model criterion and then set up the training loop. It is also instructed other details such as use of early stopping by evaluating the IoU metric over the validation data to avoid overfitting, using mixed precision training to reduce memory usage and saving the model checkpoint regularly after a given number of epochs. Specific instructions are also provided to ensure that certain commonly observed errors are eliminated.

Prompt 5
Training Loop

- Setup:
  - Specify the Adam optimizer with a learning rate of 2e-4.
  - Set up the criterion for the model training using the loss function created previously, with a weight of 0.7 towards the BCE loss.
- Train loop:
  - Set up training and validation loop in the train model function.
  - Configure early stopping based on validation loss to prevent overfitting. Allow user to set the patience for early stopping.
  - Setup verbose logging to monitor training and validation losses and IoUs per epoch. Include a tqdm progress bar for each epoch.
  - Save the train and validation losses and IoU(s) over all the epochs for plotting them after the training.
  - Use mixed precision training using GradScaler and autocast to reduce memory usage during model training.
  - Save model checkpoint after every k number of epochs. Allow user to decide the number k as an argument.
  - Save the best performing model over all epochs separately.
- Ensure the following to avoid errors:
  - Ensure that masks and outputs are in the correct format to ensure compatibility with Dice loss. Raise errors if the format and shape is incompatible. In the error mention the current output and mask shapes, their format and the required format for the compatibility.
  - Ensure that masks and outputs are in the correct format to ensure compatibility with IoU(s) computations. Raise errors if the format and shape is incompatible. In the error mention the current output and mask shapes, their format and the required format for the compatibility.
  - Apply softmax for multiclass and sigmoid for binary segmentation to the outputs, depending on the number of output channels.
  - Reshape masks and outputs wherever necessary to resolve any possible dimension errors.
  - If tensors are required to be converted to NumPy arrays, then move them first to the CPU before the conversion.

Fig. 8 Prompt to set up model training

2.5 Evaluating Model Performance on Test Data

The performance of the trained model is evaluated on the test data, again with the IoU metric, but also with help of visualizations. For this purpose, a function is created that runs one forward pass on the test data by instantiating the trained model. In the forward pass, the loss and the IoU metric over the test samples is calculated and their average values are returned. To ensure that this is executed without errors, the prompt includes a break down of the steps involved in a forward pass.

The predictions are visualized in two ways: (1) by comparing the true masks of the test samples with the predicted ones, and (2) by overlaying the predicted masks over the true masks and marking the true positive, true negative, false positive and false negative pixels with different colors. The functions for both these visualizations are generated with the instructions in the prompt shown in Fig. 9 .

3 Generated Code and Discussion

The code generated based on the prompts provided for the microstructure image segmentation is presented in this section. Along with the code listings, the outputs generated and some remarks on the generated code are also mentioned.

3.1 Loading and Pre-processing Data

From the prompts Figs. 3, 4 and 5, the following Code Listing 1 was generated. ChatGPT 4 generates a Python script that imports required libraries, creates a function that inserts randomization seeds and then a class named `MicrostructureDataset` which extracts images and masks according to the specified details, applies transformation and returns two variables `images` and `masks` which contain the extracted data.

In response to the prompt mentioned in Figs. 4 and 5, ChatGPT 4 first creates the function `get_training _augmentation`, which sets all the transformations mentioned in the prompt in the pipeline in the mentioned sequence. It also uses the appropriate function `get_preprocessing_fn` to normalize the image data with the mean and standard deviation of the 'resnet50' model architecture pre-trained on the 'imagenet' dataset. Similarly the function `get_validation _augmentation` is also created which only normalizes the image data.

After the data augmentation, the chatbot uses the `DataLoader` function from Pytorch, along with the previously created `MicrostructureDataset` to create dataloaders for training, validation and testing. Finally, a visualization function is also created to plot the augmented images and their corresponding masks.

Prompt 6
Model Testing and Visualizing Predictions

- Test model: Create a Python function named `test_model` to reliably evaluate the test performance of a segmentation model:
  - Inputs:
    - model: The trained segmentation model is set up from the weights saved during training.
    - test_loader: DataLoader for the test data.
    - criterion: The loss function is used again to evaluate the model.
    - device: The device on which the model is running (CPU or GPU).
  - Functionality:
    - Set the model to evaluation mode.
    - Initialize lists to store batch-wise test losses and IoU scores.
    - Iterate over the test data using the DataLoader.
  - For each batch:
    - Move the inputs and masks to the specified device.
    - Ensure masks are in the correct shape [batch_size, num_classes, height, width] using `.permute` if necessary.
    - Ensure masks are in the correct data type (float for criterion and long for IoU calculation).
    - Perform a forward pass through the model to get the outputs.
    - Apply `torch.softmax` for multiclass segmentation and `torch.sigmoid` for binary segmentation.
    - Compute the loss using the specified criterion.
    - Convert masks to integer type for IoU calculation.
    - Compute the IoU score using the compute_iou function.
    - Append the loss and IoU score to their respective lists.
    - Calculate the average test loss and IoU score over all batches.
    - Print and return the average test loss and IoU score.
  - Ensure robustness:
    - Move all tensors to the CPU before converting to numpy arrays for calculations.
    - Handle any potential errors related to tensor shapes and data types with informative error messages.
- Visualize Prediction: Create a Python function that plots predicted masks with the true masks for a test image for comparison. Plot them such that next to the image, we have the true masks (matrix, tertiary and secondary) and beneath them we have the corresponding predicted masks. Allow the user to decide the number of samples. Ensure that the true mask, predicted

Fig. 9 Prompt to test the trained model and visualize the predictions

mask and image dimensions are correctly handled to avoid any dimension issues in plotting.

- Visualize Prediction Accuracy: Create a Python function that overlays the true and predicted masks, and then shows the true positive pixels in white, the true negative pixels in black, the false positive pixels in blue and the false negative pixels in pink. Plot these overlayed masks, with the correspondingly colored pixels next to each image in the test samples. Include a common legend for all the colors of the pixels at the beginning of the subplots.

Use all these functions. Load the best model and evaluate the test performance with the visualization.

Fig. 9 (continued)

```
import os
import numpy as np
import cv2
import torch
import random
from torch.utils.data import Dataset, DataLoader
from albumentations import Compose, Normalize, Resize
from albumentations.pytorch import ToTensorV2

# Set random seeds for reproducibility
SEED = 42
random.seed(SEED)
np.random.seed(SEED)
torch.manual_seed(SEED)
torch.cuda.manual_seed_all(SEED)
torch.backends.cudnn.deterministic = True
torch.backends.cudnn.benchmark = False

# Load Data from directory

# Paths to dataset
base_directory = '/kaggle/input/ni-super-alloy/Super1'
train_images_dir = os.path.join(base_directory, 'train')
train_masks_dir = os.path.join(base_directory, 'train_annot')
val_images_dir = os.path.join(base_directory, 'val')
val_masks_dir = os.path.join(base_directory, 'val_annot')
test_images_dir = os.path.join(base_directory, 'test')
test_masks_dir = os.path.join(base_directory, 'test_annot')

class MicrostructureDataset(Dataset):
    def __init__(self, images_dir, masks_dir, transform=None):
        self.images_dir = images_dir
        self.masks_dir = masks_dir
```

```python
        self.transform = transform
        self.images = sorted(os.listdir(images_dir))
        self.masks = sorted(os.listdir(masks_dir))

    def __len__(self):
        return len(self.images)

    def __getitem__(self, idx):
        img_path = os.path.join(self.images_dir, self.images[idx])
        mask_path = os.path.join(self.masks_dir, self.masks[idx])

        # Read image
        image = cv2.cvtColor(cv2.imread(img_path), cv2.COLOR_BGR2RGB)

        # Read mask
        mask = cv2.imread(mask_path, 1)

        # Extract binary masks for each class
        matrix_mask = np.all(mask == [0, 0, 0], axis=-1).astype(np.uint8)
        secondary_mask = np.all(mask == [255, 0, 0], axis=-1).astype(np.uint8)
        tertiary_mask = np.all(mask == [0, 0, 255], axis=-1).astype(np.uint8)
        masks = [matrix_mask, secondary_mask, tertiary_mask]
        if mask.shape[2] > 1:
            masks[0] = ~np.any(masks[1:], axis=0)


        # Stack masks
        mask = np.stack(masks, axis=-1)

        if self.transform:
            augmented = self.transform(image=image, mask=mask)
            image = augmented['image']
            mask = augmented['mask']

        return image, mask

# Data Augmentation

import albumentations as A
from albumentations.pytorch import ToTensorV2
from segmentation_models_pytorch.encoders import get_preprocessing_params

# Get normalization parameters from segmentation_models_pytorch
preprocessing_params = get_preprocessing_params('resnet50')
mean = preprocessing_params['mean']
std = preprocessing_params['std']
```

```python

# Define training augmentations
def get_training_augmentation_pmm():
    return A.Compose([
        A.HorizontalFlip(p=0.75),
        A.RandomRotate90(p=1),
        A.GaussNoise(p=0.5),
        A.OneOf([
            A.CLAHE(),
            A.RandomBrightnessContrast(brightness_limit=0.25,
    ↪ contrast_limit=0.25),
            A.RandomGamma()
        ], p=0.5),
        A.OneOf([
            A.Sharpen(),
            A.Blur(blur_limit=3)
        ], p=0.5),
        A.OneOf([
            A.RandomBrightnessContrast(),
            A.HueSaturationValue()
        ], p=0.5),
        A.Normalize(mean=mean, std=std),
        ToTensorV2()
    ])

# Define validation augmentations
def get_validation_augmentation():
    return A.Compose([
        A.Normalize(mean=mean, std=std),
        ToTensorV2()
    ])


# Create DataLoaders and visualize some samples

import matplotlib.pyplot as plt

# Custom dataset class already defined above

# Define training, validation, and test datasets with
    ↪ augmentations
train_dataset = MicrostructureDataset(train_images_dir,
    ↪ train_masks_dir, transform=get_training_augmentation_pmm
    ↪ ())
val_dataset = MicrostructureDataset(val_images_dir,
    ↪ val_masks_dir, transform=get_validation_augmentation())
test_dataset = MicrostructureDataset(test_images_dir,
    ↪ test_masks_dir, transform=get_validation_augmentation())

# Define DataLoaders
train_loader = DataLoader(train_dataset, batch_size=4, shuffle=
    ↪ True, num_workers=0)
```

```
val_loader = DataLoader(val_dataset, batch_size=4, shuffle=
    ↪ False, num_workers=0)
test_loader = DataLoader(test_dataset, batch_size=1, shuffle=
    ↪ False, num_workers=0)

# Function to visualize augmented images and their masks
def visualize_augmentations(dataset, num_samples=4):
    fig, axs = plt.subplots(num_samples, 4, figsize=(20, 5 *
    ↪ num_samples))

    for i in range(num_samples):
        idx = random.randint(0, len(dataset) - 1)
        image, mask = dataset[idx]

        # Convert tensor to numpy array for visualization
        image_np = image.permute(1, 2, 0).cpu().numpy()
        mask_np = mask.cpu().numpy()

        # Plot image and masks
        axs[i, 0].imshow(image_np)
        axs[i, 0].set_title("Image")
        axs[i, 1].imshow(mask_np[..., 0], cmap='gray')
        axs[i, 1].set_title("Matrix Mask")
        axs[i, 2].imshow(mask_np[..., 1], cmap='gray')
        axs[i, 2].set_title("Secondary Mask")
        axs[i, 3].imshow(mask_np[..., 2], cmap='gray')
        axs[i, 3].set_title("Tertiary Mask")

        for j in range(4):
            axs[i, j].axis('off')

    plt.tight_layout()
    plt.show()

# Visualize augmentations
visualize_augmentations(train_dataset, num_samples=4)
```

Code Listing 1 Loading and pre-processing data

It is important for the user to visualize the augmented images and their masks. The augmented images should encapsulate some of the transformations in the augmentation function and should differ in terms of these transformations from each other. Moreover, the masks, 'matrix', 'secondary' and 'tertiary' should be consistent for all images. A common error observed was overlooking the instruction to update the 'matrix' mask with all the remaining unassigned pixels after extracting the 'secondary' and 'tertiary' masks. Errors in extracting the masks, and assigning them to the corresponding pixels in the image will hinder the model from training well. The major issue being the incorrect assignment of labels to the pixels, and thus obtaining a higher loss despite multiple epochs of training the model. Figure 10, shows one of the augmented images and its corresponding masks.

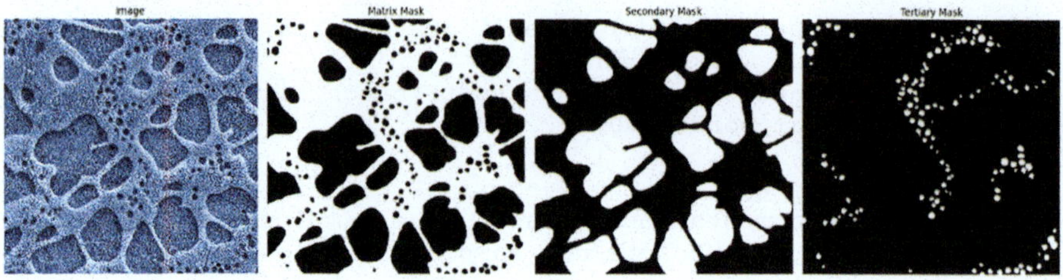

Fig. 10 Sample of augmented image and its corresponding masks

3.2 Model Setup

Using the model setup prompts Fig. 6, the functions to setup the segmentation model from the pre-trained weights, the computation of combined Dice and BCE loss, and the computation of the IoU metric were created. The function `get_pretrained_microscopynet_url` that generates the url for downloading the pre-trained weights is also mentioned here. It is directly taken as is from the GitHub repository of the reference [6] and not generated using ChatGPT 4.

```
import torch
import torch.utils.model_zoo as model_zoo
from segmentation_models_pytorch import UnetPlusPlus
from segmentation_models_pytorch.encoders import get_preprocessing_params

# Assuming get_pretrained_microscopynet_url function is already provided
# from custom_utils import get_pretrained_microscopynet_url

def get_pretrained_microscopynet_url(encoder, encoder_weights, version=1.1,
                                     self_supervision=''):
    """Get the url to download the specified pretrained encoder.

    Args:
        encoder (str): pretrained encoder model name (e.g. resnet50)
        encoder_weights (str): pretraining dataset, either 'micronet' or
            'imagenet-micronet' with the latter indicating the encoder
            was first pretrained on imagenet and then finetuned on microscopynet
        version (float): model version to use, defaults to latest.
            Current options are 1.0 or 1.1.
        self_supervision (str): self-supervision method used. If self-supervision
```

```
            was not used set to '' (which is default).

    Returns:
        str: url to download the pretrained model
    """

    # there is an error with the name for resnext101_32x8d so
    ↪ catch and return
    # (currently there is only version 1.0 for this model so
    ↪ don't need to check version.)
    if encoder == 'resnext101_32x8d':
        return 'https://nasa-public-data.s3.amazonaws.com/
    ↪ microscopy_segmentation_models/
    ↪ resnext101_pretrained_microscopynet_v1.0.pth.tar'

    # only resnet50/micronet has version 1.1 so I'm not going
    ↪ to overcomplicate this right now.
    if encoder != 'resnet50' or encoder_weights != 'micronet':
        version = 1.0

    # setup self-supervision
    if self_supervision != '':
        version = 1.0
        self_supervision = '_' + self_supervision

    # correct for name change for URL
    if encoder_weights == 'micronet':
        encoder_weights = 'microscopynet'
    elif encoder_weights == 'image-micronet':
        encoder_weights = 'imagenet-microscopynet'
    else:
        raise ValueError("encoder_weights must be 'micronet' or
    ↪  'image-micronet'")

    # get url
    url_base = 'https://nasa-public-data.s3.amazonaws.com/
    ↪ microscopy_segmentation_models/'
    url_end = '_v%s.pth.tar' %str(version)
    return url_base + f'{encoder}{self_supervision}_pretrained_
    ↪ {encoder_weights}' + url_end

# Model Setup

def setup_segmentation_model(encoder_name='resnet50',
    ↪ class_values=None, encoder_weights='micronet'):
    if class_values is None:
        raise ValueError("class_values must be provided and
    ↪ should not be None")

    # Determine number of classes
    num_classes = len(class_values)

    # Define activation function based on number of classes
```

```
    activation = 'softmax2d' if num_classes > 1 else 'sigmoid'

    # Initialize U-Net++ model
    model = UnetPlusPlus(
        encoder_name=encoder_name,
        encoder_weights=None,  # Skip default weight loading
        in_channels=3,
        classes=num_classes,
        activation=activation
    )

    # Determine device
    device = torch.device('cuda' if torch.cuda.is_available()
    ↪ else 'cpu')
    map_location = 'cuda' if torch.cuda.is_available() else '
    ↪ cpu'

    # Load custom weights
    url = get_pretrained_microscopynet_url(encoder_name,
    ↪ encoder_weights)
    state_dict = model_zoo.load_url(url, map_location=
    ↪ map_location)
    model.encoder.load_state_dict(state_dict)

    # Move model to the appropriate device
    model = model.to(device)

    return model, device

# Loss and IoU metric

import torch
import torch.nn.functional as F
import numpy as np
import segmentation_models_pytorch as smp

# Combined Dice and BCE loss function
def dice_bce_loss(inputs, targets, bce_weight=0.5):
    # Apply sigmoid to inputs
    inputs = torch.sigmoid(inputs)

    # Flatten inputs and targets
    inputs = inputs.reshape(-1) #manual
    targets = targets.reshape(-1) #manual

    # Compute Dice loss
    intersection = (inputs * targets).sum()
    dice_loss = 1 - (2. * intersection + 1) / (inputs.sum() +
    ↪ targets.sum() + 1)

    # Compute BCE loss using logits
    bce_loss = F.binary_cross_entropy_with_logits(inputs,
    ↪ targets.float())
```

```
    # Combine Dice and BCE losses
    combined_loss = dice_loss + bce_weight * bce_loss
    return combined_loss

# IoU metric function using smp
def compute_iou(output, target):
    # Get statistics for IoU calculation
    tp, fp, fn, tn = smp.metrics.get_stats(output, target, mode
    ↪ ='multilabel', threshold=0.5)

    # Compute IoU score
    iou_score = smp.metrics.iou_score(tp, fp, fn, tn, reduction
    ↪ ="micro")
    return iou_score
```

Code Listing 2 Model setup loss function and evaluation metric function

As shown in Code Listing 2, a function for computing the combined Dice and BCE loss is created. In some trials, ChatGPT 4 overlooked the specific instruction of using `torch.reshape` method to change the dimensions of the outputs and targets in the loss function. It used the `torch.Tensor.view` method, which serves the same purpose of changing array dimensions, but is not generally applicable to both contiguous and non-contiguous tensors. The IoU is computed according to the example provided in the prompt.

3.3 Training the Model

Putting all the previous functions to use, the model is trained by creating a `train_model` function according to the prompt Fig. 8. An appropriate train model function is created following the instructions in the prompt. However, the instructions for ensuring that errors are avoided are not always followed. The train model function can be then used by asking the chatbot for an example usage. The number of epochs, patience parameter for early stopping, learning rate and the weighing criteria in the loss function need to be adjusted to train the model with a higher IoU metric value. Additionally, the user can also prompt ChatGPT 4 to make a plot showing the decrease in the loss values over the number of epochs.

```
import os
import torch
import torch.optim as optim
from torch.cuda.amp import GradScaler, autocast
from torch.utils.data import DataLoader
import numpy as np
from tqdm import tqdm

# Define the training and validation loop
```

```python
def train_model(model, train_loader, val_loader, criterion, optimizer, num_epochs, patience, device, checkpoint_interval):
    best_model_wts = None
    best_iou = 0.0
    epochs_no_improve = 0

    # Store losses and IoUs for plotting
    train_losses = []
    val_losses = []
    train_ious = []
    val_ious = []

    scaler = GradScaler()

    for epoch in range(num_epochs):
        print(f'Epoch {epoch}/{num_epochs - 1}')
        print('-' * 10)

        # Each epoch has a training and validation phase
        for phase in ['train', 'val']:
            if phase == 'train':
                model.train()  # Set model to training mode
                data_loader = train_loader
            else:
                model.eval()   # Set model to evaluate mode
                data_loader = val_loader

            running_loss = 0.0
            running_iou = 0.0

            # Iterate over data with progress bar
            with tqdm(total=len(data_loader), desc=f'{phase.capitalize()} Epoch {epoch + 1}', unit='batch') as pbar:
                for inputs, masks in data_loader:
                    inputs = inputs.to(device)
                    masks = masks.to(device)
                    masks = masks.permute(0, 3, 1, 2)  # Correcting mask shape

                    # Check mask and input shapes for compatibility
                    if masks.shape != inputs.shape:
                        raise ValueError(f"Mask shape {masks.shape} and input shape {inputs.shape} are incompatible for Dice loss and IoU computation.")

                    # Zero the parameter gradients
                    optimizer.zero_grad()

                    # Forward pass
                    with autocast(enabled=True):
                        outputs = model(inputs)
```

```
                        if outputs.shape[1] == 1:
                            outputs = torch.sigmoid(outputs)
                        else:
                            outputs = torch.softmax(outputs, dim=1)

                        loss = criterion(outputs, masks)

                    if phase == 'train':
                        # Backward pass and optimization
                        scaler.scale(loss).backward()
                        scaler.step(optimizer)
                        scaler.update()

                    # Statistics
                    running_loss += loss.item() * inputs.size(0)
                    iou = compute_iou(outputs, masks)
                    running_iou += iou * inputs.size(0)

                    # Update progress bar
                    pbar.update(1)
                    pbar.set_postfix(loss=loss.item(), iou=iou.item())

            epoch_loss = running_loss / len(data_loader.dataset)
            epoch_iou = running_iou / len(data_loader.dataset)

            if phase == 'train':
                train_losses.append(epoch_loss)
                train_ious.append(epoch_iou)
            else:
                val_losses.append(epoch_loss)
                val_ious.append(epoch_iou)

            print(f'{phase} Loss: {epoch_loss:.4f} IoU: {epoch_iou:.4f}')

            # Early stopping
            if phase == 'val':
                if epoch_iou > best_iou:
                    best_iou = epoch_iou
                    best_model_wts = model.state_dict()
                    epochs_no_improve = 0
                    torch.save(model.state_dict(), 'best_model.pth')
                else:
                    epochs_no_improve += 1

        # Checkpoint the model every 'checkpoint_interval' epochs
        if (epoch + 1) % checkpoint_interval == 0:
```

```
            checkpoint_path = f'model_checkpoint_epoch_{epoch + 1}.pth'
            torch.save(model.state_dict(), checkpoint_path)
            print(f'Model checkpoint saved at {checkpoint_path}')

        if epochs_no_improve >= patience:
            print('Early stopping triggered')
            break

    # Load best model weights
    if best_model_wts:
        model.load_state_dict(best_model_wts)

    # Save losses and ious for plotting, ensuring tensors are moved to CPU
    np.save('train_losses.npy', np.array(train_losses))
    np.save('val_losses.npy', np.array(val_losses))
    np.save('train_ious.npy', np.array([iou.cpu().numpy() for iou in train_ious]))
    np.save('val_ious.npy', np.array([iou.cpu().numpy() for iou in val_ious]))

    return model

# Usage

class_values = [0, 1, 2]  # Example class values
model, device = setup_segmentation_model(encoder_name='resnet50', class_values=class_values, encoder_weights='micronet')

optimizer = optim.Adam(model.parameters(), lr=2e-4)
criterion = lambda outputs, masks: dice_bce_loss(outputs, masks, bce_weight=0.7)

# Train model
num_epochs = 250
patience = 30
checkpoint_interval = 10  # Save model every 10 epochs
model = train_model(model, train_loader, val_loader, criterion, optimizer, num_epochs, patience, device, checkpoint_interval)
```

Code Listing 3 Function to train the model

The following items must be checked with a few number of epochs (e.g. 50 epochs with a patience value of 5) before running the `train_model` function with a higher number of epochs and a high patience value for early stopping:

- Check the shapes of the masks and images are correct. Ideally, the function should raise an error if the shapes of the image and masks are inappropriate for loss and IoU metric computation.
- Check if the trained model is being saved regularly after a given interval of epochs.

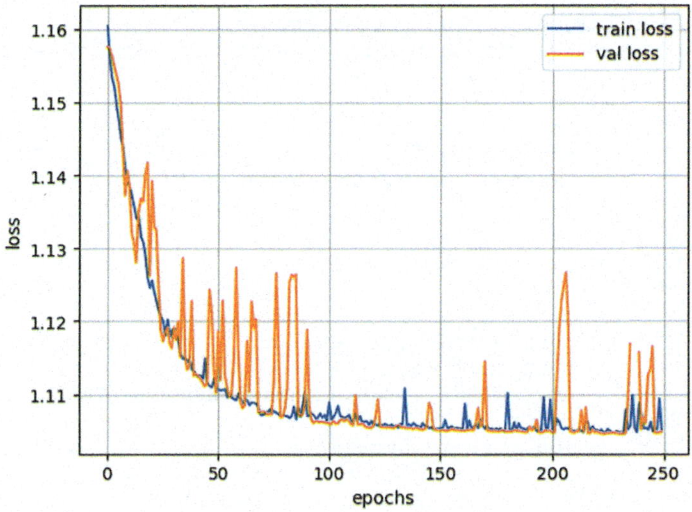

Fig. 11 Training and validation loss over epochs

- Ensure that the training and validation losses and IoU values are being stored after each epoch.
- A relatively large and quick decrease in the training loss in the first few epochs, and a consistent increase in the validation IoU value as the losses decrease.
- Check if early stopping is triggered if the validation IoU value of the epochs does not exceed the last best validation IoU value.

Using the function in Code Listing 3, the model was trained for 250 epochs and a patience of 30 epochs. The plot in Fig. 11 shows the change in training and validation loss over the epochs.

3.4 Evaluating Model Performance on Test Data

Functions for model evaluation and the required visualizations are generated by ChatGPT 4 as per the prompt Fig. 9. At times, the dimensions of the masks might not be handled correctly and in such a scenario, minor human intervention is required.

```
import matplotlib.pyplot as plt
import matplotlib.patches as mpatches

# Function to evaluate the model on the test set
def test_model(model, test_loader, criterion):
    model.eval()
    test_loss, test_iou = [], []

    with torch.no_grad():
```

```
        for images, masks in test_loader:
            images, masks = images.to(device), masks.to(device)

            # Convert masks to the shape [batch_size,
    ↪ num_classes, height, width]
            if masks.ndim == 4 and masks.shape[-1] == 3:
                masks = masks.permute(0, 3, 1, 2)  # Change
    ↪ shape from [batch_size, height, width, num_classes] to [
    ↪ batch_size, num_classes, height, width]
            masks = masks.float()  # Ensure masks are in the
    ↪ correct format

            outputs = model(images)
            if outputs.shape[1] > 1:  # Multiclass segmentation
                outputs = torch.softmax(outputs, dim=1)
            else:  # Binary segmentation
                outputs = outputs  # Use logits directly

            loss = criterion(outputs, masks)
            masks_int = masks.long()  # Convert masks to
    ↪ integer type for IoU calculation
            iou = compute_iou(outputs, masks_int)

            test_loss.append(loss.item())
            test_iou.append(iou.cpu().numpy())  # Move IoU to
    ↪ CPU for numpy operations

    avg_test_loss = np.mean(test_loss)
    avg_test_iou = np.mean(test_iou)

    print(f'Test Loss: {avg_test_loss:.4f}, Test IoU: {
    ↪ avg_test_iou:.4f}')
    return avg_test_loss, avg_test_iou

# Function to visualize predictions and true masks
def visualize_predictions(model, test_dataset, device,
    ↪ num_samples=4):
    model.eval()
    fig, axs = plt.subplots(2 * num_samples, 4, figsize=(20, 10
    ↪  * num_samples))

    for i in range(num_samples):
        idx = random.randint(0, len(test_dataset) - 1)
        image, true_mask = test_dataset[idx]
        image = image.to(device).unsqueeze(0)

        with torch.no_grad():
            output = model(image)
            if output.shape[1] == 1:
                output = torch.sigmoid(output)
            else:
                output = torch.softmax(output, dim=1)
            output = output.squeeze().cpu().numpy()
```

```python

        image = image.squeeze().permute(1, 2, 0).cpu().numpy()
        true_mask = true_mask.cpu().numpy()  # Ensure correct shape

        # Plot original image
        axs[2 * i, 0].imshow(image)
        axs[2 * i, 0].set_title("Image")

        # Plot true masks
        axs[2 * i, 1].imshow(true_mask[..., 0], cmap='gray')
        axs[2 * i, 1].set_title("True Matrix Mask")
        axs[2 * i, 2].imshow(true_mask[..., 1], cmap='gray')
        axs[2 * i, 2].set_title("True Secondary Mask")
        axs[2 * i, 3].imshow(true_mask[..., 2], cmap='gray')
        axs[2 * i, 3].set_title("True Tertiary Mask")

        # Plot predicted masks
        axs[2 * i + 1, 0].imshow(image)
        axs[2 * i + 1, 0].set_title("Image")
        axs[2 * i + 1, 1].imshow(output[0], cmap='gray')
        axs[2 * i + 1, 1].set_title("Predicted Matrix Mask")
        axs[2 * i + 1, 2].imshow(output[1], cmap='gray')
        axs[2 * i + 1, 2].set_title("Predicted Secondary Mask")
        axs[2 * i + 1, 3].imshow(output[2], cmap='gray')
        axs[2 * i + 1, 3].set_title("Predicted Tertiary Mask")

        for j in range(4):
            axs[2 * i, j].axis('off')
            axs[2 * i + 1, j].axis('off')

    plt.tight_layout()
    plt.show()

# Function to overlay predicted and true masks, highlighting errors
def visualize_prediction_accuracy(model, test_dataset, device, num_samples=4):
    model.eval()
    fig, axs = plt.subplots(num_samples, 4, figsize=(20, 5 * num_samples))

    for i in range(num_samples):
        idx = random.randint(0, len(test_dataset) - 1)
        image, true_mask = test_dataset[idx]
        image = image.to(device).unsqueeze(0)

        with torch.no_grad():
            output = model(image)
            if output.shape[1] == 1:
                output = torch.sigmoid(output)
            else:
                output = torch.softmax(output, dim=1)
```

```python
            output = output.squeeze().permute(1, 2, 0).cpu().numpy()


        image = image.squeeze().permute(1, 2, 0).cpu().numpy()
        true_mask = true_mask.cpu().numpy()

        #print("output: ", output.shape)
        #print("image: ", image.shape)
        #print("true mask: ", true_mask.shape)
        if output.shape != true_mask.shape:
            raise ValueError(f"Shape mismatch: Predicted mask shape {output.shape} and true mask shape {true_mask.shape} are incompatible for visualization.")

        axs[i, 0].imshow(image)
        axs[i, 0].set_title("Image")

        for j in range(3):
            pred_mask = output[:, :, j] > 0.5
            true_class_mask = true_mask[:, :, j]

            overlay = np.zeros((*true_class_mask.shape, 3), dtype=np.uint8)
            #print("pred_mask :", pred_mask.shape)
            #print("true_class_Mask: ", true_class_mask.shape)

            true_positives = (pred_mask == 1) & (true_class_mask == 1)
            true_negatives = (pred_mask == 0) & (true_class_mask == 0)
            false_positives = (pred_mask == 1) & (true_class_mask == 0)
            false_negatives = (pred_mask == 0) & (true_class_mask == 1)

            overlay[true_positives] = [255, 255, 255]  # White
            overlay[true_negatives] = [0, 0, 0]        # Black
            overlay[false_positives] = [0, 0, 255]     # Blue
            overlay[false_negatives] = [255, 105, 180] # Pink

            axs[i, j + 1].imshow(overlay)
            axs[i, j + 1].set_title(f"Mask {j+1} Overlay")

        for j in range(4):
            axs[i, j].axis('off')

    white_patch = mpatches.Patch(color='white', label='True Positive')
    black_patch = mpatches.Patch(color='black', label='True Negative')
    blue_patch = mpatches.Patch(color='blue', label='False Positive')
```

```
    pink_patch = mpatches.Patch(color='pink', label='False Negative')
    plt.legend(handles=[white_patch, black_patch, blue_patch, pink_patch], loc='upper right')
    plt.tight_layout()
    plt.show()

# Usage

# 250 epochs, patience  = 30, never reached early stopping
model.load_state_dict(torch.load('model_checkpoint_epoch_250.pth'))

# Evaluate the model on the test set
test_loss, test_iou = evaluate_model(model, test_loader, criterion, device)

    # Visualize predictions and accuracy
#visualize_predictions(model, test_dataset, device, num_samples=4)
visualize_prediction_accuracy(model, test_dataset, device, num_samples=4)
```

Code Listing 4 Functions for evaluating model performance on test data and visualizing predictions

The `test_model` function runs a forward pass by using the trained model with the test dataset and computes the loss and the IoU metric. While this function posed no errors in trials, it is still suggested to check if the test accuracy and IoU value are close to the ones observed for the epoch where the best model was saved. If there is a significant deviation, the first step is to check if the `test_model` function has been implemented correctly. If the deviation is significant despite having a correctly implemented function, then there are possible issues in the training of the model. However, such errors did not arise during any of the trials.

Besides checking the loss and IoU values, a visual observation can also provide a good idea if the model is performing well. Here, the visualization functions are useful. In Fig. 12a and b, the predicted masks and the true masks for two models: one trained with 50 epochs and the other trained with 250 epochs, can be observed.

The prediction accuracy can also be checked by overlaying the predicted masks over the true masks and marking the false positive and false negative pixels. The Fig. 13a and b show the for the model trained for 50 epochs and the one trained for 250 epochs.

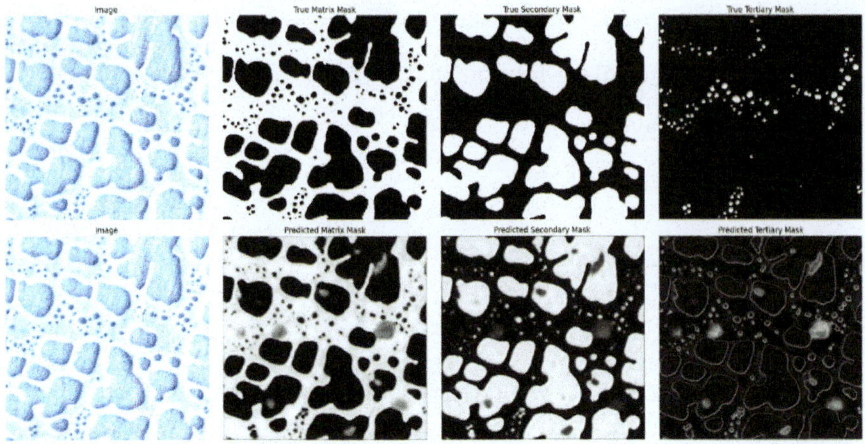

(a) 50 epochs

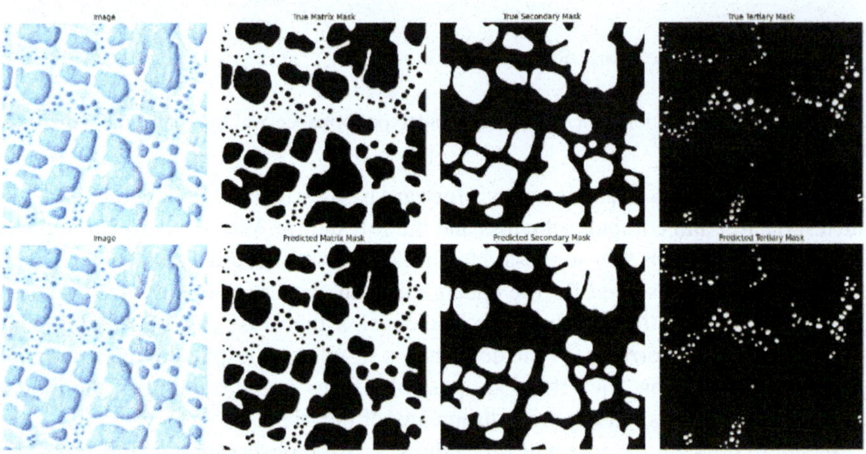

(b) 250 epochs

Fig. 12 True and predicted masks for a model trained for different numbers of epochs. The first row in **a** and **b** each displays the true masks and the second row in **a** and **b** show the predicted masks for the same image

4 Discussion

Microstructure image segmentation using transfer learning is a relatively complex task, primarily due to the involvement of multiple sub-steps that each require the usage of different Python libraries and functions. Moreover, since most machine learning tasks do not have a definite solution, there is a large room for possible solution approaches. This intrinsic freedom of machine learning approaches in such complex tasks demands multiple iterations and variations, which can be

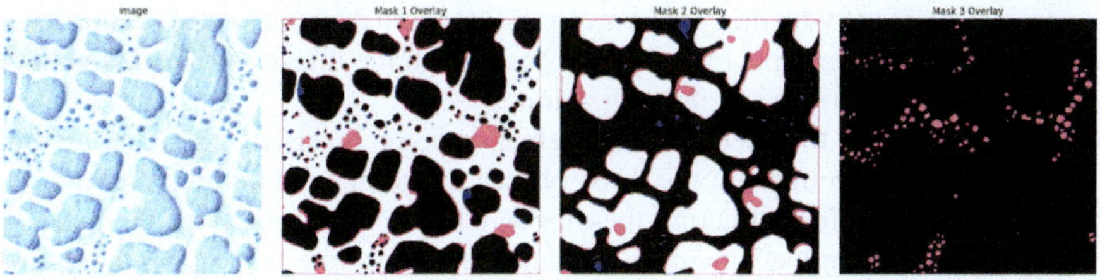

(a) 50 epochs

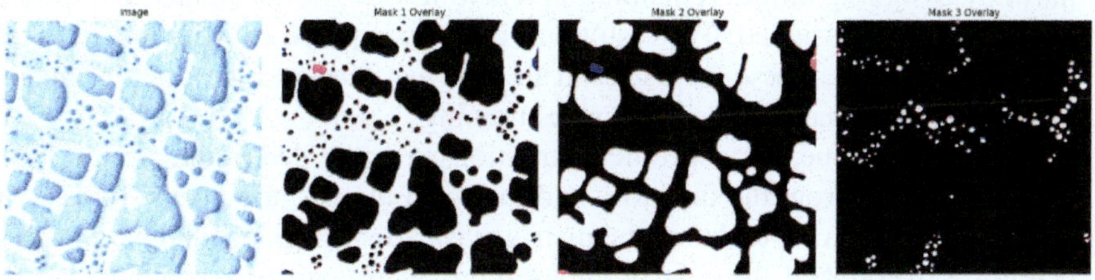

(b) 250 epochs

Fig. 13 Prediction accuracy visualization for the model trained for different number of epochs. False negative pixels are colored pink and false positive pixels are colored blue

hugely aided by LLMs like ChatGPT 4. However, if the solution approach is already decided, then the LLM should be instructed with details and required specifics to expect desirable outcomes. Here, the solution approach is as per the example notebook `multiclass_segmentation_example.ipynb` provided with [6] is followed, and thus any room for assumptions must be eliminated by providing detailed instructions. With appropriate prompts, it can be observed that ChatGPT 4 provides suitable Python codes for executing the task. The generated codes had rare instances of syntax errors, and a few logical errors. In this section, these errors shall be discussed. In addition, the changes made in the prompts and directly to the generated codes to avoid these errors are also included. The errors are further categorized as 'major errors' and 'minor errors'.

4.1 Major Errors

- **Extraction of images**: If specific instructions on the commands for loading the images from the directories are not provided, ChatGPT 4 changes the image dimensions to the ones appropriate for common deep learning architectures like 'ResNet50'. It also augments the images with random transformations and normalizes them with inappropriate mean and standard deviation values.
  For the user to ensure that the correct transformations are applied to the images, the user must specify the required steps in the augmentation pipeline and their

parameters, along with the correct mean and standard deviation values to normalize the images. The prompts Figs. 3 and 4 ensure that the images are loaded with the desired augmentations and normalization.

- **Assigning masks to pixels**: In the original data, the pixels are annotated in three different colors to indicate the three phases in the microstructure. Without specific instructions for extracting the masks, ChatGPT 4 converts all the masks to grayscale and then loads them to the data for performing the segmentation task. Furthermore, ChatGPT 4 often overlooked the instruction to update the matrix mask by assigning all the remaining pixels to it after the secondary and tertiary masks had been assigned. The result was the matrix phase constituting all pixels which were black. These errors result in incorrect labels to the sample images, which shall eventually hinder the model training.
  As a fix to this issue, very specific instructions are provided in the prompt Fig. 3 to extract masks from the directory and to assign them to the pixels. In case the specific instruction for updating the matrix mask after assigning all other masks to pixels was overlooked, the same prompt was provided again to the chatbot. Usually, after providing the same prompt again, all the specific details are considered in the code generation.
- **Sticking to old versions of imported library methods**: For some specific cases, e.g. for using the IoU metric from the `segmentation_models_pytorch` library, ChatGPT 4 used the syntax from the older versions. Until the latest syntax of the functions was not provided explicitly through an example usage from the library documentation, ChatGPT 4 continued to use the older version. This might be possibly due to the fact that LLMs such as ChatGPT 4 are not trained frequently, and the current version might be trained when the older versions of these libraries existed.
  By just providing an example usage in prompt Fig. 6, ChatGPT 4 could adapt its code response to the latest syntax of the IoU metric.
- **Model checkpointing**: Despite specific instructions for model checkpointing in prompt Fig. 8, ChatGPT 4 missed including regular model checkpointing after a given number of epochs. This step is crucial if the training takes places over a high number of epochs and requires long training periods.

4.2 Minor Errors

- **Dimension errors**: ChatGPT 4 produces syntactically correct code. However, it is quite frequent to get dimension errors, mainly in the dimensions of the masks. The underlying reason might be that ChatGPT 4 may not be effective in tracking the changes in the dimensions of the masks when they are passed through multiple functions throughout the data loading, pre-processing and training process.
  An attempt to mitigate this error was to explicitly request ChatGPT 4 to ensure that there are no dimension errors in the code it generates. This can be observed

in the system prompt Fig. 2. However, despite such an instruction, the code had some cases of dimension error. Another important step to reduce the number of dimension errors, is to specify the shape of the image and mask files in the prompt. This reduces the frequency of dimension errors, especially in loading the data. Finally, ChatGPT 4 is instructed to raise errors if the dimensions are inappropriate, for e.g. in the `train_model` function, before passing the masks and the outputs to the loss functions.

- **Moving variables to CPU before typecasting to NumPy arrays**: Since the model is allowed to work on the GPU for training, the PyTorch tensors and the model are all mapped to the GPU. However, when certain functions or visualizations require the variables as NumPy arrays, the PyTorch tensors are not always moved to the CPU before converting them to NumPy arrays.
  By providing specific instructions in the prompts, e.g. in prompts Figs. 2, 8 and 9, this error could be eliminated.
- **Reshaping PyTorch tensor dimensions**: Two methods are commonly used to reshape PyTorch tensors: `torch.reshape` and `torch.Tensor.view`. The latter is not applicable to non-contiguous tensors, i.e. it is not generally applicable. This posed errors in the loss function for the combined Dice and BCE losses. ChatGPT 4 often overlooked the instruction to use `torch.reshape`.

5 Conclusion

Segmentation of microstructure images into the classes 'matrix', 'secondary' and 'tertiary' using a ResNet50 encoder pre-trained on the MicroNet dataset and a UNet++ decoder was conducted using ChatGPT 4. The key steps in accomplishing this were discussed and their corresponding codes were generated by providing ChatGPT 4 with carefully drafted prompts. ChatGPT 4 provided promising results in generating the code for such a complex task. Certain major and minor errors were observed in the process, and their fixes were discussed likewise. An iterative process was required to obtain the desired output since in machine learning tasks, there is no definite solution, and thus the chatbot has a large room for making its own assumptions, if details are not provided. Here, the errors encountered in these iterations could be tackled by adding specific instructions to the prompt. It was also observed that GPT-4 overlooked some specific instructions of the prompts. The missed-out instructions were taken into consideration by ChatGPT 4 after providing it with the same prompt. Similar prompts as mentioned for this task can be applied to binary or multi-class segmentation depending on the specific microstructures. Furthermore, using transfer learning makes this approach robust for few-shot learning, i.e. for training the model with a very small number of labeled microstructure images. As a result, the predicted masks can be used to estimate at high accuracy the amount of phases present in a given microstructure image and further analyses.

References

1. Akers, S., Kautz, E., Trevino-Gavito, A., Olszta, M., Matthews, B., Wang, L., Du, Y. & Spurgeon, S. Rapid and flexible segmentation of electron microscopy data using few-shot machine learning. *Npj Computational Materials*. **7**, 187 (2021,11,17), https://doi.org/10.1038/s41524-021-00652-z
2. Stan, T., Thompson, Z. & Voorhees, P. Building towards a universal neural network to segment large materials science imaging datasets. *Developments In X-Ray Tomography XII*. **11113** pp. 111131G (2019), https://doi.org/10.1117/12.2525290
3. Zhuang, F., Qi, Z., Duan, K., Xi, D., Zhu, Y., Zhu, H., Xiong, H. & He, Q. A Comprehensive Survey on Transfer Learning. (2020)
4. He, K., Zhang, X., Ren, S. & Sun, J. Deep Residual Learning for Image Recognition. (2015)
5. Zhou, Z., Siddiquee, M., Tajbakhsh, N. & Liang, J. UNet++: A Nested U-Net Architecture for Medical Image Segmentation. (2018)
6. Stuckner, J., Harder, B. & Smith, T. Microstructure segmentation with deep learning encoders pre-trained on a large microscopy dataset. *Npj Computational Materials*. **8**, 200 (2022,9,19), https://doi.org/10.1038/s41524-022-00878-5

Topics of Computational Analysis of Waves and Fluid Mechanics

Elastic Wave Propagation

Muhammad Saad Qureshi and Bernhard Eidel

Abstract This chapter investigates the ability of ChatGPT to generate code in the field of Computational Materials Science, specifically using GPT-4. The focus of this study is on solving the wave equation in 2D, considering elastic wave propagation, and visualizing the results using Python with a finite difference time domain (FDTD) scheme. The wave equation is fundamental in understanding how waves, such as sound or seismic waves, travel through different media. The FDTD method is a numerical technique used to model wave propagation by discretizing both time and space. This involves setting up a computational grid, applying initial and boundary conditions, and iteratively solving the wave equation to simulate the wave's behavior over time. Visualization of the results helps in interpreting the physical phenomena and verifying the accuracy of the computational model.

1 Introduction

Artificial intelligence, particularly in the form of Large Language Models (LLMs) like ChatGPT has shown significant potential in various fields including code generation; for a pioneering work focusing on numerical methods see [1]. This chapter explores ChatGPT's capabilities for generating a Python code that solves the problem of 2D elastic wave propagation based on the Finite Difference Time Domain (FDTD) method. In particular, this work aims at verifying the code, assessing its accuracy and efficiency in simulating and visualizing the behavior of wave propagation. Overall, we highlight both the strengths and limitations of GPT-4 in handling complex scientific problems.

M. S. Qureshi · B. Eidel (✉)
M5, IMFD, TUBAF, Freiberg, Germany
e-mail: bernhard.eidel@imfd.tu-freiberg.de

M. S. Qureshi
e-mail: muhammad-saad.qureshi@student.tu-freiberg.de

© The Author(s), under exclusive license to Springer Nature Switzerland AG 2025

B. Eidel (ed.), *GPT for Python-Coding in Computational Materials Science and Mechanics*, Studies in Computational Intelligence 1198,
https://doi.org/10.1007/978-3-031-85470-5_8

1.1 Wave Equation in 2D

The wave equation is a second-order linear partial differential equation (PDE) describing the behavior of waves. In two dimensional (x, y) coordinate system, it is defined as [4]

$$\frac{\partial^2 \boldsymbol{u}}{\partial t^2} = c^2 \left(\frac{\partial^2 \boldsymbol{u}}{\partial x^2} + \frac{\partial^2 \boldsymbol{u}}{\partial y^2} \right). \tag{1}$$

Considering elastic wave propagation in a square domain, x and y represent the axes having equal length N, $\boldsymbol{u}$ represents the displacement vector from rest position, and c represents the wave speed, calculated by $c = \sqrt{E/\rho}$, where E is the Young's modulus, and ρ is the mass density.

1.2 Finite Difference Time Domain (FDTD)

Finite difference methods are numerical techniques used to approximate the derivatives using finite differences. FDTD method was introduced by Kane S. Yee for numerically solving time-dependent Maxwell's equations [2]. The method involves discretizing the space and time partial derivatives based on central-difference approximations [5]. Applying FDTD approximations to (1) we obtain exemplarily for the derivatives of displacement component u_x

$$\frac{\partial^2 u_x}{\partial x^2} \approx \frac{u_x(x+h, y; t) - 2u_x(x, y; t) + u_x(x-h, y; t)}{h^2}, \tag{2}$$

$$\frac{\partial^2 u_x}{\partial y^2} \approx \frac{u_x(x, y+h; t) - 2u_x(x, y; t) + u_x(x, y-h; t)}{h^2}, \tag{3}$$

$$\frac{\partial^2 u_x}{\partial t^2} \approx \frac{u_x(x, y; t+\Delta t) - 2u_x(x, y; t) + u_x(x, y; t-\Delta t)}{\Delta t^2}. \tag{4}$$

Without loss of generality, we consider a square domain with a uniform spatial grid of size h both in x and y directions and for uniform temporal discretization Δt as the constant time step size. Using (2), (3), (4) in (1) along with the notations $u(x, y; t) =: u^n_{x,y}$, $u(x, y; t \pm \Delta t) =: u^{n\pm1}_{x,y}$, $u(x \pm h, y; t) =: u^n_{x\pm1,y}$, $u(x, y \pm h; t) =: u^n_{x,y\pm1}$, where n refers to the current time t_n, and $n \pm 1$ to time $t_n \pm \Delta t$, one obtains

$$\frac{1}{\Delta t^2} \left(u^{n+1}_{x,y} - 2u^n_{x,y} + u^{n-1}_{x,y} \right) = \frac{c^2}{h^2} \left(u^n_{x+1,y} + u^n_{x-1,y} + u^n_{x,y+1} + u^n_{x,y-1} - 4u^n_{x,y} \right). \tag{5}$$

Solving for $\boldsymbol{u}_{x,y}^{t+1}$ based on t_n-data yields the final form of the 2D wave equation discretized by FDTD

$$u_{x,y}^{n+1} = \frac{c^2 \Delta t^2}{h^2} \left(u_{x+1,y}^n + u_{x-1,y}^n + u_{x,y+1}^n + u_{x,y-1}^n - 4u_{x,y}^n\right) + 2u_{x,y}^n - u_{x,y}^{n-1}. \quad (6)$$

Furthermore, initial conditions (ICs), boundary conditions (BCs) have to be defined for a full-fledged Initial Boundary Values Problem (IBVP); the numerical solution scheme requires the choice of the discretization parameters Δx, Δy and Δt.

1.3 Initial Conditions

Two types of excitation functions are considered to be used for defining ICs in the domain.

- Sinusoidal function

$$u(t) = A \sin(2\pi f t + \phi), \quad (7)$$

  where A is the amplitude, f is the frequency and ϕ is the phase angle, see Fig. 1 (bottom).
- Gaussian function

$$u(t) = a \exp\left(-\frac{(t-b)^2}{2c^2}\right), \quad (8)$$

  where a is the height of the curve's peak, b is the center of the peak, and c is the standard deviation, see Fig. 1 (top).

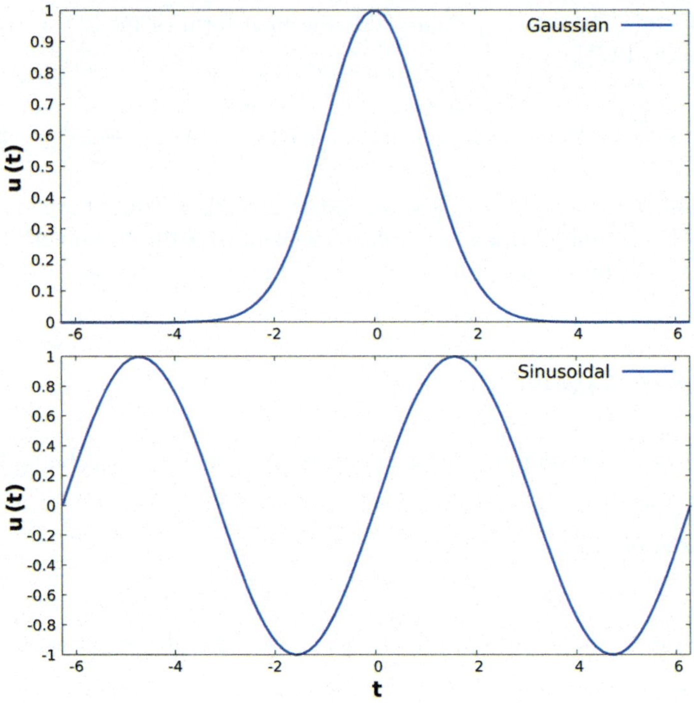

Fig. 1 Gaussian function with $a = 1$, $b = 0$, $c = 1$, and sinusoidal function with $A = 1$, $f = \frac{1}{2\pi}$, $\phi = 0$

1.4 Boundary Conditions

Two types of BCs are considered:

- Dirichlet or reflecting BC : no displacement at the boundaries, total reflection of the waves

$$u_{0,y} = u_{x,0} = u_{N,y} = u_{x,N} = 0. \tag{9}$$

- Mur or absorbing BC, first introduced for time-dependent Maxwell equations [3]: total absorption at the boundaries, no reflection of the waves. In the discretized setting, the Mur BC are [7]

$$u_{0,y}^{n+1} = u_{1,y}^{n} - \frac{h - c\Delta t}{h + c\Delta t}\left(u_{1,y}^{n+1} - u_{0,y}^{n}\right), \tag{10}$$

$$u_{N,y}^{n+1} = u_{N-1,y}^{n} + \frac{h - c\Delta t}{h + c\Delta t}\left(u_{N,y}^{n} - u_{N-1,y}^{n+1}\right), \tag{11}$$

$$u_{x,0}^{n+1} = u_{x,1}^{n} - \frac{h - c\Delta t}{h + c\Delta t}\left(u_{x,1}^{n+1} - u_{x,0}^{n}\right), \tag{12}$$

$$u_{x,N}^{n+1} = u_{x,N-1}^{n} + \frac{h - c\Delta t}{h + c\Delta t}\left(u_{x,N}^{n} - u_{x,N-1}^{n+1}\right). \tag{13}$$

Here, (10) refers to waves travelling in $-x$ direction or the left side of the square domain, (11) refers to waves travelling in $+x$ direction or the right side of the square domain, (12) refers to waves travelling in $-y$ direction or the bottom side of the square domain, (13) refers to waves travelling in $+y$ direction or the top side of the square domain.

1.5 Courant-Friedrichs-Lewy Condition

For numerical solutions of PDEs involving explicit time integration, the convergence condition of Courant-Friedrichs-Lewy (CFL) is a necessary condition for convergence. As a consequence of the explicit time integrator FDTD, the time step must be less than a certain upper bound, given a fixed spatial increment; here, with c_x and c_y the wave velocities in x and y directions, the CFL condition for a uniform spatial grid size h can be given according to [6]

$$\frac{c_x \Delta t}{h} + \frac{c_y \Delta t}{h} \leq C_{\max}, \tag{14}$$

which is met by a value for $C_{\max}$ of 1. The equality condition in (14) is exploited to calculate the critical time step $\Delta t_{\max}$. To satisfy the CFL condition, the time step Δt must be either equal or smaller than $\Delta t_{\max}$.

1.6 Strains and Stresses

Once the updated displacement field is computed for the entire domain, elastic strains are calculated using the gradient of the displacement field, and, based on that, stresses using Hooke's law. Isotropic, linear elastic material behavior is assumed to hold.

$$\varepsilon_x = \frac{\partial u_x}{\partial x}, \qquad \varepsilon_y = \frac{\partial u_y}{\partial y}, \qquad \gamma_{xy} = \frac{\partial u_y}{\partial x} + \frac{\partial u_x}{\partial y}. \tag{15}$$

Here, ε_x and ε_y are the normal strain components, and γ_{xy} is the shear strain.

$$\sigma_{xx} = \frac{E}{1 - \nu^2}(\varepsilon_x + \nu\varepsilon_y), \tag{16}$$

$$\sigma_{yy} = \frac{E}{1 - \nu^2}(\varepsilon_y + \nu\varepsilon_x), \tag{17}$$

$$\sigma_{xy} = \frac{E}{1 + \nu}(\gamma_{xy}). \tag{18}$$

Here, σ_{xx} and σ_{yy} are the normal stresses, and σ_{xy} is the shear stress. E is the Young's modulus and ν is the Poisson's ratio of the material. From the normal stresses and the shear stress, principal stresses are calculated according to

$$\sigma_1, \sigma_2 = \frac{\sigma_{xx} + \sigma_{yy}}{2} \pm \sqrt{\left(\frac{\sigma_{xx} - \sigma_{yy}}{2}\right)^2 + \sigma_{xy}^2}. \tag{19}$$

The algorithm for FDTD simulations of elastic wave propagation in 2D is given in the Algorithm Box 1.

Algorithm 1: Elastic Wave Propagation and Stress Calculation using FDTD

Input : Grid size N, grid spacing h, material properties (E_1, ν_1, ρ_1) for first half, (E_2, ν_2, ρ_2) for second half

Output: Stress distribution over time

Initialize grid size and time step based on CFL condition;
Initialize displacement fields $u^n_{x,y}$, $u^{n-1}_{x,y}$, $u^{n+1}_{x,y}$ to zero;
Initialize stress fields $\sigma_{xx}, \sigma_{yy}, \sigma_{xy}$ to zero;
Define sources with positions, frequencies, and amplitudes;
for *each time step t* **do**
    for *each grid point* (x, y) *except boundaries* **do**
        Get material properties (E, ν, c) based on position ;
        Update displacement $u^{n+1}_{x,y}$ using FDTD equation;
    Apply source excitation at the source location;
    if *Boundary condition is 'Mur'* **then**
        Apply Mur boundary conditions considering wave speed;
    else if *Boundary condition is 'Dirichlet'* **then**
        Apply Dirichlet boundary conditions (zero displacement);
    Update previous and current displacement fields;
    Compute strains $\varepsilon_x, \varepsilon_y, \gamma_{xy}$ as gradients of the displacement field;
    for *each grid point* (x, y) *except boundaries* **do**
        Calculate stresses $\sigma_{xx}, \sigma_{yy}, \sigma_{xy}$ using material properties and strains;
    if *Calculate principal stresses* **then**
        Calculate principal stresses σ_1, σ_2;
for *each time step t* **do**
    if *Calculate principal stresses* **then**
        Update the plots with σ_1, σ_2;
    else
        Update the plots with $\sigma_{xx}, \sigma_{yy}, \sigma_{xy}$;
    Render the animation frame;

1.7 Problems

The proper functionality of the generated code for the wave equation model and the FDTD solution scheme shall be tested in two problems. The problem defined in Table 1 has a corresponding reference code available on the internet [7], hence all the parameters and values are selected to match the reference code. Here, the time step size is significantly lower than the critical time step. Since this problem merely aims at a quantitative, numerical analysis of wave propagation, we can safely restrict for the parameters on their dimensions instead of explicit units. The output from the GPT-4's code and the reference code is compared and discussed in Sect. 4.

Table 1 Problem 1: Settings for the displacement study

Parameters	Values/Types	Dimensions
Domain size (N)	200	$[L]$
Domain type	Homogeneous	
Wave speed (c)	0.2	$[LT^{-1}]$
Spatial step size (h)	1	$[L]$
Critical time step size ($\Delta t_{\max}$)	2.5	$[T]$
Time step size (Δt)	1	$[T]$
BC	Dirichlet and Mur	
Type of excitation function	Sinusoidal	
Amplitude (A)	80	$[L]$
Frequency	1/50	$[T^{-1}]$
Source point	(100, 100)	$[L]$
Output	Displacement animation	

The problem in Table 2 adds further complexity to the previous problem in terms of a heterogeneous domain, two types of excitation functions, strain and stress calculations and two types of stress plots. Here, the time step size is equal to the critical time step.

Table 2 Problem 2: Settings for the stress study

Parameters	Values/Types	Units
Domain size (N)	200	mm
Domain type	Heterogeneous	
Upper half material	Aluminum	
Lower half material	Chromium	
Young's modulus of chromium	280	GPa
Poisson's ratio of chromium	0.2	
Density of chromium	7100	kg/m^3
Young's modulus of aluminium	70	GPa
Poisson's ratio of aluminum	0.33	
Density of aluminum	2700	kg/m^3
Spatial step size (h)	1	mm
Critical time step size (Δt_{max})	7.96e-5	s
Time step size (Δt)	Equal to Δt_{max}	s
BC	Dirichlet and Mur	
Type of excitation function	Sinusoidal or Gaussian	
Amplitude (A)	80	mm
Frequency	1/50	Hz
Source point	(100, 100)	mm
Output	Stress animation	

2 Prompt

The final prompt is given in Fig. 2. It is used for generating the Code Listing 1. The proper labelling of the plots had to be done manually.

For Problem 2: Stress Study, the level of flexibility aimed could not be captured in a single prompt. The Code Listing 2 is generated iteratively. In addition to that, manual intervention is carried out to label the plots.

Consider yourself an expert in computational physics and finite difference method. I need a Python script to simulate wave propagation in a two-dimensional domain using the finite difference time-domain (FDTD) method. The domain should be 200 units in size, with spatial discretization parameters (h) set at 1.0, and the time step (Δt) is 1 unit, do not use CFL condition in this case.
Make sure a consistent coordinate system is chosen. Use np.linspace and np.meshgrid to generate the mesh.
The simulation should include: Material and Wave Properties: A single homogeneous material with a wave speed of 0.2 units. A wave source located at position (100, 100) within the domain, emitting a sinusoidal wave with a frequency of 0.04 Hz and an amplitude of 80 units.
Boundary Conditions: The ability to toggle between 'Mur' (absorbing) and 'Dirichlet' (reflecting) boundary conditions: Mur conditions should minimize reflections at the boundaries, correctly calculating the transition for outgoing waves based on the local wave speed. For example, this would be the equation for -x direction:

$$u_{0,j}^{n+1} = u_{1,j}^{n} - \frac{h - c\Delta t}{h + c\Delta t}(u_{1,j}^{n+1} - u_{0,j}^{n}).$$

Dirichlet conditions should set the displacement to zero at all boundaries, effectively reflecting incoming waves back into the domain.
FDTD Update Mechanism: An update function that recalculates the wave field throughout the domain at each time step, adjusting the wave field based on both the wave source and the selected boundary condition.
Visualization: An animation that shows the evolution of the wave field over time, updating each frame to reflect changes in the wave propagation. The animation should visualize the displacement field with color mapping to represent displacement magnitude.
Additional Specifications: Ensure that the script is fully runnable with comments explaining key sections of the code, including how BCs are applied and how the wave source influences the domain.
Output: The script should allow for easy toggling between BCs and include code to start an animation visualizing the wave propagation. Optionally, provide a way to save the animation as a video file.
The structure of the program should look like this: Define parameters, grid, and initialize fields. Function 'source' that returns the source excitation. Function 'update' that involves the FDTD method and the BCs. Function 'animate' for the animation.

Fig. 2 Prompt for Problem 1: displacement study

Please provide complete Python code that meets these specifications, ensuring that all components function as described and are suitable for demonstrating basic principles of wave propagation in a simulated environment. Make sure all the variables in the code are consistent.

Fig. 2 (continued)

3 Code Listing

```
import numpy as np
import matplotlib.pyplot as plt
from matplotlib.animation import FuncAnimation

# Simulation parameters
size = 200          # domain size
dx = dy = 1.0       # spatial discretization
dt = 1.0            # time step
c = 0.2             # wave speed

# Create a spatial grid
x = np.linspace(0, size-1, size)
y = np.linspace(0, size-1, size)
X, Y = np.meshgrid(x, y)

# Initialize fields
u = np.zeros((size, size))    # current wave field
u_prev = np.zeros((size, size))  # previous wave field
u_next = np.zeros((size, size))  # next wave field

# Source parameters
source_amplitude = 80
source_frequency = 0.04
source_position = (100, 100)

def source(t):
    """ Returns the source value at time t """
    return source_amplitude * np.sin(2 * np.pi * source_frequency
    ↪  * t)

def update(u, u_prev, t, boundary_condition='Mur'):
    """ Update wave field for the next time step using the FDTD
    ↪ method """
    # Core update (finite difference)
    for i in range(1, size-1):
        for j in range(1, size-1):
            u_next[i, j] = (c**2 * dt**2 / dx**2) * (u[i+1, j] +
    ↪ u[i-1, j] + u[i, j+1] + u[i, j-1] - 4*u[i, j]) + 2*u[i, j]
    ↪  - u_prev[i, j]
```

```
    # Apply source
    u_next[source_position] += source(t)

    # Apply boundary conditions
    if boundary_condition == 'Mur':  # Absorbing boundary
        for j in range(size):
            # Left and right boundaries
            u_next[0, j] = u[1, j] - (dx-c*dt)/(dx+c*dt) * (u_next[1, j] - u[0, j])
            u_next[-1, j] = u[-2, j] - (dx-c*dt)/(dx+c*dt) * (u_next[-2, j] - u[-1, j])
            # Top and bottom boundaries
            u_next[j, 0] = u[j, 1] - (dx-c*dt)/(dx+c*dt) * (u_next[j, 1] - u[j, 0])
            u_next[j, -1] = u[j, -2] - (dx-c*dt)/(dx+c*dt) * (u_next[j, -2] - u[j, -1])
    elif boundary_condition == 'Dirichlet':  # Reflecting boundary
        u_next[0, :] = u_next[-1, :] = u_next[:, 0] = u_next[:, -1] = 0

    # Update arrays for next iteration
    u_prev[:, :] = u[:, :]
    u[:, :] = u_next[:, :]

def animate(i):
    """ Animation function """
    update(u, u_prev, i, boundary_condition='Mur')  # Toggle 'Mur' or 'Dirichlet'
    img.set_data(u)
    img.set_clim(-source_amplitude, source_amplitude)
    return img,

# Setup plot for animation
fig, ax = plt.subplots()
img = ax.imshow(u, origin='lower', cmap='viridis',
                interpolation='nearest', animated=True, vmin=-source_amplitude, vmax=source_amplitude)
ax.set_title("Wave Propagation in homogeneous medium", fontsize = 12)
ax.set_xlabel("x-axis", fontsize = 12)
ax.set_ylabel("y-axis", fontsize = 12)
cbar = fig.colorbar(img, ax=ax)
cbar.set_label('Wave Amplitude')

# Create animation
ani = FuncAnimation(fig, animate, frames=1000, interval=50, blit=True)

# Optionally save the animation
#ani.save('wave_propagation2.mp4', writer='ffmpeg')
plt.show()
```

Code Listing 1 Output by ChatGPT for solving Problem 1: Displacement Study

```
# -*- coding: utf-8 -*-
"""
Created on Fri Jul 12 19:15:36 2024

@author: Saad Qureshi
"""

import numpy as np
import matplotlib.pyplot as plt
from matplotlib import animation, rc

# Constants and Material Properties
E1 = 280e9  # Young's modulus in Pascals for first half (Chromium
    ↪ )
nu1 = 0.2   # Poisson's ratio for first half
rho1 = 7100 # Density in kg/m^3 for first half
E2 = 70e9   # Young's modulus in Pascals for second half (
    ↪ Aluminum)
nu2 = 0.33  # Poisson's ratio for second half
rho2 = 2700 # Density in kg/m^3 for second half
c1 = np.sqrt(E1 / rho1)  # Wave speed for first half
c2 = np.sqrt(E2 / rho2)  # Wave speed for second half

domain_size = 200
dx = 1.0
dy = 1.0
dt = dx / (max(c1, c2)*(2))  # CFL condition for stability

calculate_principal_stresses = True
boundary_type = 'Mur'  # Can be 'Mur' or 'Dirichlet'
excitation_type = 'gaussian'  # Can be 'sinusoidal' or 'gaussian'

# Initialize fields
x = np.linspace(0, domain_size-1, domain_size)
y = np.linspace(0, domain_size-1, domain_size)
u = np.zeros((domain_size, domain_size))
u_prev = np.zeros((domain_size, domain_size))
u_next = np.zeros((domain_size, domain_size))
stress_xx = np.zeros((domain_size, domain_size))
stress_yy = np.zeros((domain_size, domain_size))
stress_xy = np.zeros((domain_size, domain_size))

# Source definitions
sources = [{"position": (100, 100), "frequency": 0.04, "amplitude
    ↪ ": 80}]

def source(t, frequency, amplitude, excitation_type):

        if excitation_type == 'sinusoidal':
            omega = 2 * np.pi * frequency
            return amplitude * np.sin(omega * t)
```

```python
        elif excitation_type == 'gaussian':
            return amplitude * np.exp(-((t - 5) ** 2) / (2 * (0.5 ** 2)))

# Function to get properties based on location
def get_properties(i):
    if i < domain_size // 2:
        return E1, nu1, c1
    else:
        return E2, nu2, c2

# Update function with Mur boundary conditions adjusted per wave speed
def update(t):
    global u, u_prev, u_next, stress_xx, stress_yy, stress_xy
    # Apply dynamic updates within the domain
    for i in range(1, domain_size-1):
        for j in range(1, domain_size-1):
            E, nu, c = get_properties(i)
            u_next[i, j] = (2 * u[i, j] - u_prev[i, j] +
                            c**2 * dt**2 / dx**2 * (u[i+1, j] + u[i-1, j] + u[i, j+1] + u[i, j-1] - 4 * u[i, j]))

    # Apply sources
    for source_info in sources:
        src_x, src_y = source_info["position"]
        u_next[src_x, src_y] += source(t, source_info["frequency"], source_info["amplitude"], excitation_type)

    # Apply Mur boundary conditions correctly for wave speed
    if boundary_type == 'Mur':
        for j in range(1, domain_size-1):
            c_left = get_properties(1)[2]
            c_right = get_properties(domain_size-2)[2]
            u_next[0, j] = u_prev[1, j] + (c_left * dt - dx) / (c_left * dt + dx) * (u_next[1, j] - u_prev[0, j])
            u_next[-1, j] = u_prev[-2, j] + (c_right * dt - dx) / (c_right * dt + dx) * (u_next[-2, j] - u_prev[-1, j])

        for i in range(1, domain_size-1):
            c_top = get_properties(i)[2]
            c_bottom = get_properties(i)[2]
            u_next[i, 0] = u_prev[i, 1] + (c_bottom * dt - dy) / (c_bottom * dt + dy) * (u_next[i, 1] - u_prev[i, 0])
            u_next[i, -1] = u_prev[i, -2] + (c_top * dt - dy) / (c_top * dt + dy) * (u_next[i, -2] - u_prev[i, -1])

    elif boundary_type == 'Dirichlet':
        u_next[0, :] = 0
        u_next[-1, :] = 0
        u_next[:, 0] = 0
        u_next[:, -1] = 0
```

```
    u_prev, u = u, u_next.copy()

    # Recalculate stresses
    grad_u_x = np.gradient(u, axis=0)  # Gradient along x-axis
    grad_u_y = np.gradient(u, axis=1)  # Gradient along y-axis
    for i in range(1, domain_size-1):
        E, nu, _ = get_properties(i)
        stress_xx[i, 1:-1] = E / (1 - nu**2) * (grad_u_x[i, 1:-1]
    ↪   + nu * grad_u_y[i, 1:-1])
        stress_yy[i, 1:-1] = E / (1 - nu**2) * (grad_u_y[i, 1:-1]
    ↪   + nu * grad_u_x[i, 1:-1])
        stress_xy[i, 1:-1] = E / (2 * (1 + nu)) * (grad_u_x[i,
    ↪  1:-1] + grad_u_y[i, 1:-1])

    if calculate_principal_stresses:
        # Calculate principal stresses
        sigma_avg = (stress_xx + stress_yy) / 2
        sigma_diff = (stress_xx - stress_yy) / 2
        R = np.sqrt(sigma_diff**2 + stress_xy**2)
        principal_stress_1 = sigma_avg + R
        principal_stress_2 = sigma_avg - R
        return principal_stress_1, principal_stress_2
    else:
        return stress_xx, stress_yy, stress_xy

# Define the animation function and plot setup
# Define the animation function and plot setup
fig, ax = plt.subplots(1, 3, figsize=(18, 6))

# Initial settings for stress_xx plot
im1 = ax[0].imshow(stress_xx, origin='lower', extent=[0,
    ↪  domain_size * dx, 0, domain_size * dy], cmap='viridis',
    ↪  vmin = np.min(stress_xx), vmax = np.max(stress_xx))
cbar1 = fig.colorbar(im1, ax=ax[0])
cbar1.set_label('Stress')
ax[0].set_title('Stress 1')
ax[0].set_xlabel('x axis (mm)')
ax[0].set_ylabel('y axis (mm)')

# Initial settings for stress_yy plot
im2 = ax[1].imshow(stress_yy, origin='lower', extent=[0,
    ↪  domain_size * dx, 0, domain_size * dy], cmap='viridis',
    ↪  vmin = np.min(stress_yy), vmax = np.max(stress_yy))
cbar2 = fig.colorbar(im2, ax=ax[1])
cbar2.set_label('Stress')
ax[1].set_title('Stress 2')
ax[1].set_xlabel('x axis (mm)')
ax[1].set_ylabel('y axis (mm)')

# Initial settings for stress_xy plot
im3 = ax[2].imshow(stress_xy, origin='lower', extent=[0,
    ↪  domain_size * dx, 0, domain_size * dy], cmap='viridis',
    ↪  vmin = np.min(stress_xy), vmax = np.max(stress_xy))
```

```
cbar3 = fig.colorbar(im3, ax=ax[2])
cbar3.set_label('Stress')
ax[2].set_title('Stress 3')
ax[2].set_xlabel('x axis (mm)')
ax[2].set_ylabel('y axis(mm)')

# Main title for the figure
fig.suptitle('Stress Distribution Over Time', fontsize=16)

# Function to update plots
def animate(t):
    if calculate_principal_stresses:
        stress_1, stress_2 = update(t)
        im1.set_data(stress_1)
        im2.set_data(stress_2)
        im3.set_data(np.zeros_like(stress_1))  # Placeholder if
    ↪ only two plots needed
    else:
        stress_xx, stress_yy, stress_xy = update(t)
        im1.set_data(stress_xx)
        im2.set_data(stress_yy)
        im3.set_data(stress_xy)
    return im1, im2, im3

ani = animation.FuncAnimation(fig, animate, frames=200, interval
    ↪ =50, blit=True)
plt.tight_layout()
ani.save('stress_ani2.mp4', writer='ffmpeg')
plt.show()
```

Code Listing 2 Output by ChatGPT for solving Problem 2: Stress Study

4 Tests for Verification

For the displacement study in Problem 1, a reference code [7] –equally based on the FDTD method– is used to test the output of GPT-4's code. The problem parameters as listed in Table 1 are chosen exactly the same for a proper comparison.

4.1 Test 1

The output is compared for both types of BCs.

Figure 3 indicates total reflection for reflection BCs. Vice versa, Fig. 4 indicates total absorption for absorbing BCs. Furthermore, the same pattern emerge for both the cases. Hence, it is concluded that the ICs, BCs and the FDTD method are working as intended.

Fig. 3 Snapshot of the amplitude animation in case of reflecting BCs

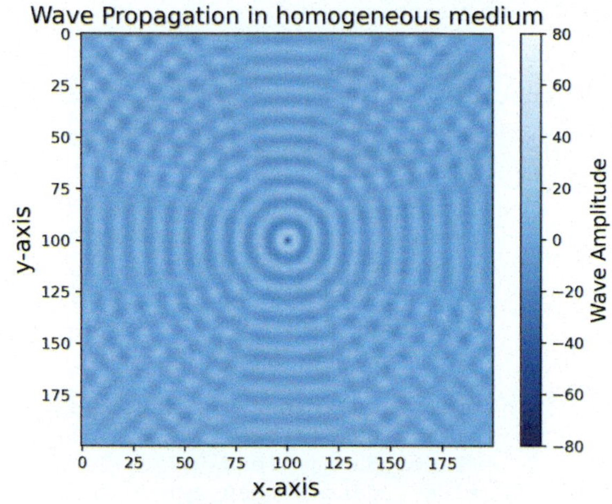

(a) Reference code

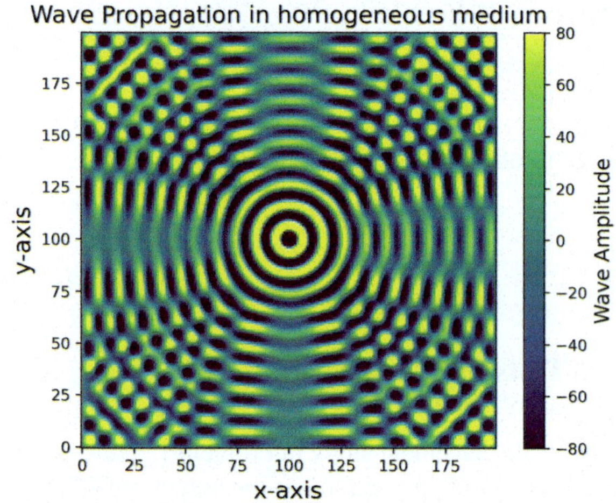

(b) GPT-4 code

Fig. 4 Snapshot of the amplitude animation in case of absorbing BCs

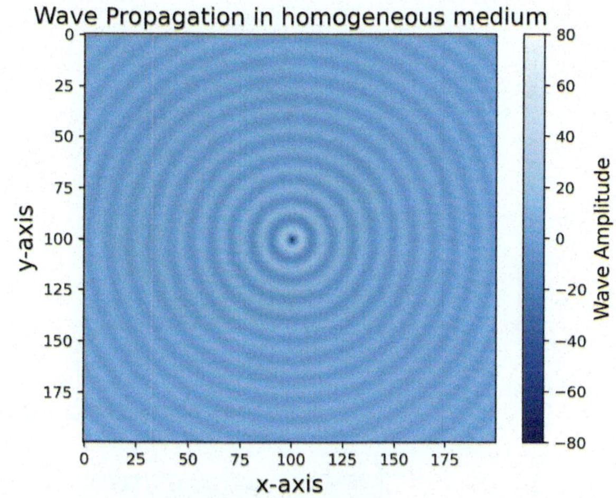

(a) Reference code

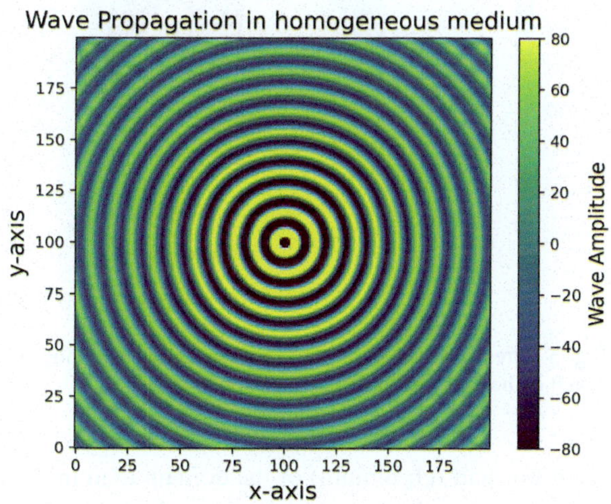

(b) GPT-4 code

4.2 Test 2

For the stress study in Problem 2, the solution shall be tested for plausibility for a layered domain in terms of stresses and strains. Note that there are no units associated with the colormap. In the code, the minimum and maximum of colormap are defined as the minimum and maximum of the respective stress, but due to the dynamic nature of the plot, the program is unable to provide determine a useful range of the colormap. From the Code Listing 8.2, the correct calculation of strain and stress are easily verified. Furthermore, Fig. 5 indicates that the stress field travels faster in

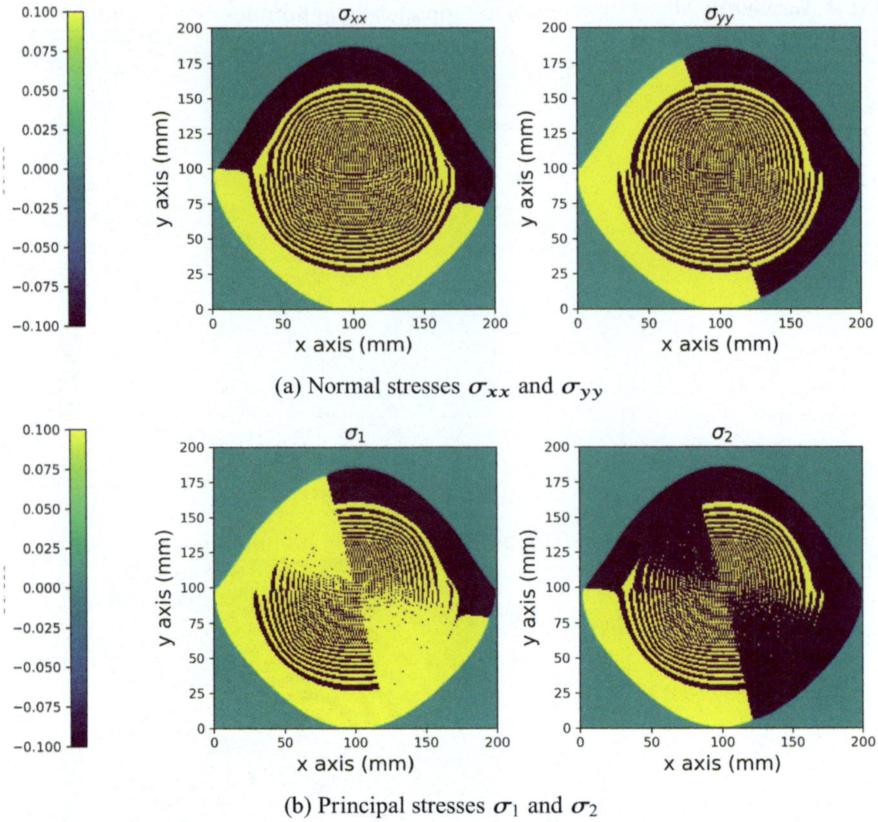

Fig. 5 Stress contour plots for Gaussian excitation along with Mur BCs and a higher speed of sound in the lower half of the simulation domain

the lower half (chromium) of the domain than in the upper half (aluminum), which is consistent with chromium having a higher wave speed constant, 6280 m/s, than aluminum, 5090 m/s.

These conclusions can further be applied to the other excitation function and BC.

5 Discussion

The methodology to get the desired output is:

- Start out with a detailed prompt, covering all the features required in the code.
- Test the code. In case of errors, convey the errors to GPT-4 and what should be the correct output. Keep doing these iterations until the code with correct output is generated.

- Finally modify the prompt with the knowledge gained from the iterations. Ask GPT-4 to specially focus in areas where most frequent errors occurred. Briefly explain the structure of the required code in terms of different functions etc., using the knowledge of the code with correct output.

The methodology mentioned above works for the displacement study of Problem 1, but does not work properly for the stress study in a heterogeneous domain referred to as Problem 2; the reason is the multitude of features that are requested seemingly can not be successfully captured in one prompt in a way that it is processed by the Chatbot.

For that reason, the set of Problem 2 calls for a solution strategy of prompt engineering, where all the required features in the code are communicated in several prompts. All attempts to generalize these successive prompts have failed, since they resulted in a large variety of different code outputs generated by GPT-4 in each try. It is this kind of instability to generate for the same prompt a unique code of granted quality, which forbids to present a prompt for Problem 2 in this chapter. Instead we restrict to the generated code that, finally, solved the problem.

Some of the important aspects that are noticed while working on the prompts, generating code and testing are:

- GPT-4 frequently uses variables that were not defined in the code. Hence, a NameError shows up while running the code.
- When a detailed prompt is provided to GPT-4, it sometimes ignores certain features completely. Sometimes it defines a certain function only as a comment line describing its usage, and the implementation part is left empty.
- Another common error of GPT-4 is not to define the origin of the domain correctly. The x-axis of the plot would start from the bottom left corner of the square and the y-axis from the top left corner of the square.
- GPT-4 makes a lot of errors regarding the animation part of the code. A lot of debugging and iterations are required to make that part of the code function as intended.
- Even if an expression is explicitly defined, GPT-4 would sometimes use a different one. For example, CFL formula to calculate Δt would be explained in the prompt, but GPT-4 would use a different formula.
- A recurring issue is GPT-4 coming to a halt while generating code. The speed of the text generation is very slow.
- Reproducibility remains a severe issue for GPT-4. While testing the prompt mentioned in Code Listing 8.1, the correct output is generated most of the times, but not always. In case there are some errors in the output code, regenerating the code usually results in removal of those errors.
- GPT-4 makes a lot of errors while implementing absorbing BC, which would result in reflection of waves at the boundaries. To fix this, the discretized formula (10) is included in the prompt as a reference, which resolved the issue.

6 Conclusion

GPT-4 was used to numerically solve the 2D wave equation using finite difference time domain scheme. As it turned out, GPT-4 exhibits enough domain expertise both for the wave problem and the numerical method to generate a correct code solution for a wave propagation displacement study along with absorbing as well as reflecting boundary conditions. In a problem of wave propagation in a heterogeneous medium along with stress computation, hence a task of increased complexity, GPT-4 revealed deficits which called for manual/human intervention. Despite its demonstrated performance in code generation, GPT-4 showed its current limitations in this project; it suffers (i) from unstable code generation, where exactly the same prompt results in different outputs, and (ii) from the inability to account for all the different tasks in a prompt for complex problems. These disadvantages make formulating single prompts for complex problems difficult, and make successive prompts and manual intervention unavoidable.

References

1. Kashefi, A., Mukerji, T. ChatGPT for programming numerical methods. *Journal of Machine Learning for Modeling and Computing*. **4** pp. 1–74 (2023), http://dx.doi.org/10.1615/JMachLearnModelComput.2023048492
2. Yee, K. Numerical solution of initial boundary value problems involving maxwell's equations in isotropic media. *Institute of Electrical and Electronics Engineers (IEEE)*. **14** pp. 302–307 (1966), http://dx.doi.org/10.1109/TAP.1966.1138693
3. Mur, G. Absorbing Boundary Conditions for the Finite-Difference Approximation of the Time-Domain Electromagnetic-Field Equations. *IEEE Transactions on Electromagnetic Compatibility*. **EMC-23** pp. 377–382 (1981), http://dx.doi.org/10.1109/TEMC.1981.303970
4. Agarwal, R.P., O'Regan, D. Two-Dimensional Wave Equation. In: Ordinary and Partial Differential Equations. *Springer, New York, NY.* pp. 292–299 (2009), https://doi.org/10.1007/978-0-387-79146-3_37
5. Linge, S., Langtangen, H.P. Wave Equations. In: Finite Difference Computing with PDEs. *Texts in Computational Science and Engineering, vol 16. Springer, Cham.* pp. 93–205 (2017), https://doi.org/10.1007/978-3-319-55456-3_2
6. Wikipedia contributors. (https://en.wikipedia.org/wiki/Courant-Friedrichs-Lewy_condition), [Online; accessed July-2024]
7. Hill, C. (https://scipython.com/blog/the-two-dimensional-wave-equation/), [Online; accessed July-2024]

Electromagnetic Wave Propagation in Dielectric Media

Tejas Viresh Anvekar and Bernhard Eidel

Abstract This study leverages the Conditional Generative Pre-trained Transformer, specifically ChatGPT-4, to develop a Python-based application for analyzing the propagation of electromagnetic waves in dielectric media using the two-dimensional Finite-Difference Time-Domain (FDTD) approach. The primary aim is to examine how electromagnetic waves interact with various dielectric environments, focusing on their reflection, transmission, and absorption properties. This work aims to exploit the capabilities of ChatGPT-4 to craft an accurate simulation tool. The performance of the generated FDTD simulations by ChatGPT-4 is evaluated. The findings suggest that ChatGPT-4 successfully creates FDTD program codes that conform to expected physical outcomes, albeit with slight variances. This research underscores the potential and accuracy of AI in handling sophisticated electromagnetic simulation tasks.

1 Introduction

The study of electromagnetic wave propagation through various media is essential for developments in telecommunications, radar, and electronic systems engineering. The Finite-Difference Time-Domain (FDTD) method, pioneered by Yee in 1966, offers a dynamic numerical technique for solving Maxwell's equations in the time-domain. This method is highly regarded for its direct algorithmic form, which is capable of managing complex boundary conditions (BC) and varying media properties.

With the advancement of computational methods, FDTD has been widely applied in conjunction with fluid dynamics concepts to enhance the study of wave propagation in various media. Fluid dynamics principles, such as those encapsulated by the Boltzmann equation, allow for the modeling of wave interactions in complex

T. V. Anvekar · B. Eidel (✉)
M5, IMFD, TUBAF, Freiberg, Germany
e-mail: bernhard.eidel@imfd.tu-freiberg.de

T. V. Anvekar
e-mail: Tejas-Viresh.Anvekar@student.tu-freiberg.de

© The Author(s), under exclusive license to Springer Nature Switzerland AG 2025
B. Eidel (ed.), *GPT for Python-Coding in Computational Materials Science and Mechanics*, Studies in Computational Intelligence 1198,
https://doi.org/10.1007/978-3-031-85470-5_9

media. By coupling fluid dynamic behavior (through the distribution of particles or fields) with electromagnetic wave propagation, we gain a deeper understanding of the interaction of waves with different materials. The Boltzmann-FDTD method, which bridges the dynamics of particles with wave behavior, captures essential characteristics such as reflection, refraction, and absorption. This approach, grounded in fluid dynamics, enhances the accuracy of simulations by incorporating the underlying physical properties of the medium, offering insights into energy transfer, wave dynamics, and material interactions.

With the advent of artificial intelligence (AI), particularly advancements in machine learning and algorithm generation, new possibilities have emerged to enhance and automate numerical simulations. The Conditional Generative Pretrained Transformer, ChatGPT-4, by OpenAI, is a testament to these advancements, providing sophisticated language comprehension and code generation capabilities. This work employs ChatGPT-4 to devise a Python program that simulates electromagnetic wave interactions in dielectric media using the FDTD method.

The primary goal of this research is to create a comprehensive simulation platform to study how electromagnetic waves, specifically Gaussian pulses and plane waves, interact within different dielectric environments. These wave types are chosen for their relevance in practical applications and their distinct propagation characteristics, which are crucial for a detailed study of wave phenomena including reflection, transmission, absorption, and scattering.

One of the key challenges in simulating electromagnetic wave propagation is choosing the appropriate dimensionality of the model. While 3D simulations offer the most detailed analysis, they are computationally expensive and time-consuming. To strike a balance between accuracy and computational efficiency, this work transitions from a 3D model to a 2D cross-sectional simulation. The 2D model captures essential phenomena such as reflection, refraction, and absorption while significantly reducing the computational burden. This approach is illustrated in Fig. 1, where the 3D simulation box with a spherical dielectric medium is simplified to a 2D cross-sectional view.

The dielectric medium used in this study is central to the investigation of electromagnetic wave behavior. The medium is characterized by a specific relative permittivity and conductivity, which are chosen to allow detailed observation of critical wave phenomena. The dielectric cylinder placed at the center of the simulation domain, as shown in Fig. 1, serves as a test object to visualize reflection, refraction, and absorption of electromagnetic waves. These phenomena are not only fundamental to understanding wave-material interactions but are also crucial for the development of practical devices in communications and radar systems.

This work also explores various BC, particularly the Perfectly Matched Layer (PML), to minimize reflections at the boundaries of the computational domain, which is vital for accurate simulation of open-region electromagnetic problems. By integrating ChatGPT-4 in the development process, this study seeks to merge computational electromagnetics with AI, enabling rapid prototyping and validation of electromagnetic models. The simulations produced by ChatGPT-4 will be rigorously compared with analytical models and established FDTD software outputs to assess their accu-

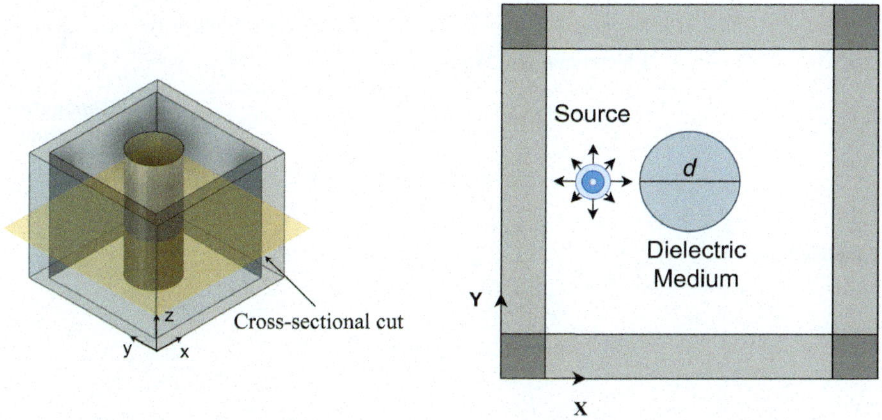

(a) 3D Simulation Box (b) 2D Cross-Sectional View

Fig. 1 Transition from 3D to 2D simulation for efficient analysis. The left image **a** shows a 3D simulation setup, while the right image **b** illustrates the 2D cross-sectional view used for the simulation with a cylindrical dielectric medium with a diameter (d) of 20 cm. The dielectric cylinder is positioned at the center of the domain. The Gaussian source emits in all directions

racy and effectiveness. This investigation not only evaluates the reliability of AI in generating viable scientific computing algorithms but also examines its potential to simplify complex simulation tasks [1, 6, 7].

1.1 Electro-Magnetic Theory

Maxwell's equations form the foundation of classical electromagnetism, describing how electric and magnetic fields propagate and interact with matter. In a dielectric medium, these equations can be expressed in a differential form as follows [1]:

$$\nabla \cdot \boldsymbol{D} = \rho, \tag{1}$$

$$\nabla \cdot \boldsymbol{B} = 0, \tag{2}$$

$$\nabla \times \boldsymbol{E} = -\frac{\partial \boldsymbol{B}}{\partial t}, \tag{3}$$

$$\nabla \times \boldsymbol{H} = \boldsymbol{J} + \frac{\partial \boldsymbol{D}}{\partial t}, \tag{4}$$

where $\boldsymbol{E}$ is the electric field, $\boldsymbol{H}$ is the magnetic field, $\boldsymbol{D}$ is the electric flux density, $\boldsymbol{B}$ is the magnetic flux density, ρ is the charge density, and $\boldsymbol{J}$ is the current density. These equations can be discretized for computational purposes using the FDTD method.

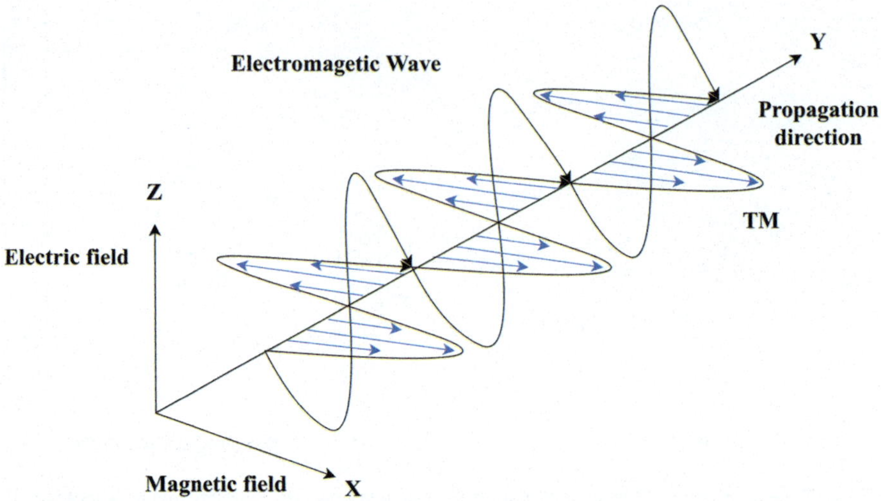

Fig. 2 Illustration of a Transverse Magnetic (TM) wave. The magnetic field is confined to the x-y plane, while the electric field has a component in the z direction

In a 2D FDTD simulation, we consider the TM (Transverse Magnetic) waves, where the electric field has only a z-component (E_z), and the magnetic fields have x and y components (H_x, H_y). In this mode, the electric field E_z is oriented along the z-axis, while the magnetic field components H_x and H_y are oriented in the x and y directions, respectively, and are transverse to the direction of wave propagation as visualized in Fig. 2. The Maxwell's curl equations for this mode are

$$\frac{\partial H_x}{\partial t} = -\frac{1}{\mu}\frac{\partial E_z}{\partial y}, \tag{5}$$

$$\frac{\partial H_y}{\partial t} = \frac{1}{\mu}\frac{\partial E_z}{\partial x}, \tag{6}$$

$$\frac{\partial E_z}{\partial t} = \frac{1}{\epsilon}\left(\frac{\partial H_y}{\partial x} - \frac{\partial H_x}{\partial y}\right), \tag{7}$$

where ϵ is the permittivity and μ is the permeability of the medium. These equations are discretized in both time and space, allowing for the fields to be updated iteratively at each time step.

1.2 Dimensionality Reduction

To address computational complexity and memory requirements in FDTD simulations, it is standard practice to implement these simulations on a 2-D lattice. In this work, we assume that the field components are independent of the z coordinate ($\partial/\partial z = 0$). Under this assumption, the plane electromagnetic field can be decomposed into transverse electric (TE) and transverse magnetic (TM) waves. These wave modes are characterized by the following relations [3]:

$$\text{TE}: \; E_z = 0, \quad H_x = H_y = 0 \tag{8}$$

$$\text{TM}: \; H_z = 0, \quad E_x = E_y = 0 \tag{9}$$

Considering the TM wave, the update equation for the electric field component E_z is

$$E_z^{n+1}(i, j) = E_z^n(i, j) + \frac{\Delta t}{\epsilon}\left[\frac{H_y^n(i, j) - H_y^n(i-1, j)}{\Delta x} - \frac{H_x^n(i, j) - H_x^n(i, j-1)}{\Delta y}\right]. \tag{10}$$

For the magnetic field components H_x and H_y the update equations are

$$H_x^{n+1/2}(i, j) = H_x^{n-1/2}(i, j) - \frac{\Delta t}{\mu}\frac{E_z^n(i, j+1) - E_z^n(i, j)}{\Delta y}, \tag{11}$$

$$H_y^{n+1/2}(i, j) = H_y^{n-1/2}(i, j) + \frac{\Delta t}{\mu}\frac{E_z^n(i+1, j) - E_z^n(i, j)}{\Delta x}. \tag{12}$$

Here, Δx and Δy represent the spatial steps in the x and y directions, respectively, while Δt denotes the time step size. The indices i and j refer to the grid points in space, and n indicates the time step. The magnetic field components H_x and H_y are staggered with respect to the electric field component E_z. This staggering is crucial for accurately calculating the fields at each time step, following the Yee cell structure, which is a standard approach in FDTD simulations. The interleaving of these fields, as depicted in Fig. 3, ensures that the electromagnetic fields propagate correctly throughout the computational domain. These update equations are applied iteratively to achieve this propagation.

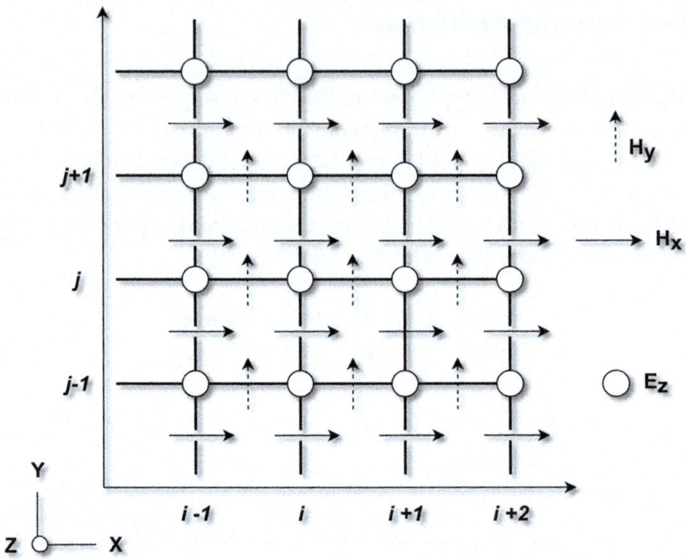

Fig. 3 Interleaving of the ***E*** and ***H*** fields for the two-dimensional TM formulation, sketch adopted from [4]

1.3 Absorbing Boundary Conditions (BC)

In FDTD simulations, accurate boundary conditions (BC) are essential to minimize reflections from the edges of the computational domain, which can interfere with the results. One of the most effective methods for absorbing outgoing waves is the perfectly matched layer (PML) BC.

The PML is designed to absorb electromagnetic waves at the boundary of the simulation domain, preventing them from reflecting back into the interior. This is achieved by gradually increasing the material's conductivity in the PML region, which attenuates the wave as it propagates through the layer. The key to the effectiveness of PML is that it matches the impedance of the interior domain, ensuring minimal reflection.

1.3.1 Updating with PML

The implementation of PML in a 2D FDTD simulation involves modifying the update equations for the electric and magnetic fields in the PML region. The electric and magnetic field components are split into two auxiliary variables, which are updated separately. Maxwell's equations can be updated as follows [5]

$$D_z^{n+1/2}(i,j) = g_{i3}(i)g_{j3}(j)D_z^{n-1/2}(i,j) + 0.5g_{i2}(i)g_{j2}(j) \times \left[H_y^n\left(i+\frac{1}{2},j\right) - H_y^n\left(i-\frac{1}{2},j\right) - H_x^n\left(i,j+\frac{1}{2}\right) + H_x^n\left(i,j-\frac{1}{2}\right)\right]. \tag{13}$$

As illustrated in Fig. 4, the parameter distribution within the PML ensures that outgoing waves are effectively absorbed. The overlapping regions at the corners indicate where both sets of parameters intersect, which is crucial for minimizing reflections. The parameters $g_{i2}, g_{i3}, g_{j2}, g_{j3}$ are given by

$$g_{i2} = g_{j2} = \frac{1}{1 + \dfrac{\sigma \Delta t}{2\epsilon_0}} \tag{14}$$

$$g_{i3} = g_{j3} = \frac{\dfrac{\sigma \Delta t}{2\epsilon_0}}{1 + \dfrac{\sigma \Delta t}{2\epsilon_0}} \tag{15}$$

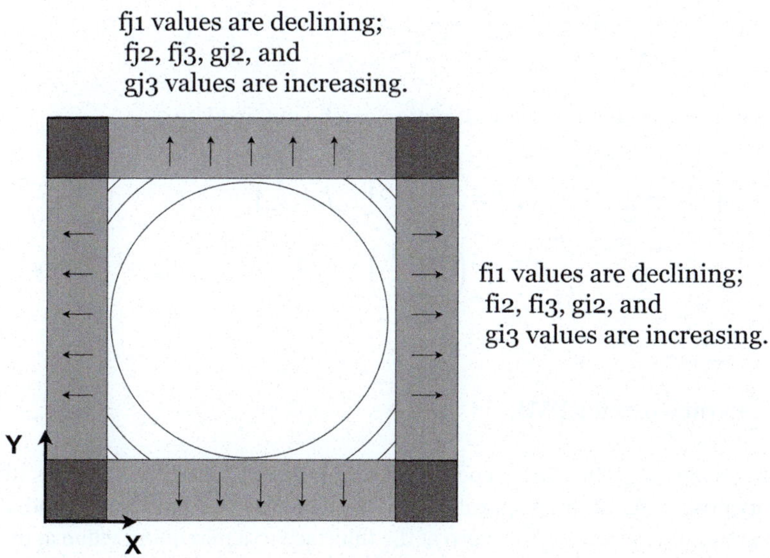

Fig. 4 This figure illustrates the parameter distribution within the PML (Perfectly Matched Layer) region of a 2D FDTD simulation. The overlapping regions at the corners indicate where both sets of parameters intersect, ensuring effective absorption of outgoing waves

The curl of the electric field can be given by

$$\nabla \times \boldsymbol{E} = E_x^{n+1/2}(i+1, j) - E_x^{n+1/2}(i, j). \tag{16}$$

The incident magnetic field in the y-direction can be given by

$$H_y^{n+1/2}\left(i+\frac{1}{2}, j\right) = H_y^{n-1/2}\left(i+\frac{1}{2}, j\right) + \nabla \times \boldsymbol{E}. \tag{17}$$

The total magnetic field in the y-direction can be evaluated recursively as follows

$$\begin{aligned} H_y^{n+1/2}\left(i+\frac{1}{2}, j\right) &= g_{i3}\left(i+\frac{1}{2}\right) H_y^n\left(i+\frac{1}{2}, j\right) \\ &\quad - 0.5 g_{i2}\left(i+\frac{1}{2}\right) \nabla \times \boldsymbol{E} - g_{i1}(j) H_y^n\left(i+\frac{1}{2}, j\right), \end{aligned} \tag{18}$$

where g_{i1} is given by:

$$g_{i1} = \frac{\sigma \Delta t}{2\epsilon_0}. \tag{19}$$

The incident magnetic field in the x-direction is given by

$$I_{H_x}^{n+1/2}\left(i, j+\frac{1}{2}\right) = I_{H_x}^{n-1/2}\left(i, j+\frac{1}{2}\right) + \nabla \times \boldsymbol{E}. \tag{20}$$

The total magnetic field in the x-direction can be obtained by

$$\begin{aligned} H_x^{n+1}\left(i, j+\frac{1}{2}\right) &= g_{j3}\left(j+\frac{1}{2}\right) H_x^n\left(i, j+\frac{1}{2}\right) \\ &\quad + 0.5\, g_{j2}\left(j+\frac{1}{2}\right) \nabla \times \boldsymbol{E} + g_{j1}(j) I_{H_y}^{n+1/2}\left(i, j+\frac{1}{2}\right). \end{aligned} \tag{21}$$

1.3.2 Parameters for PML

The effectiveness of the PML depends on the choice of parameters such as the conductivity profile. A common approach is to use a polynomial grading of conductivity, which gradually increases from zero at the interface to a maximum value at the outer edge of the PML. The conductivity σ in the PML can be expressed as [4]

$$\sigma(x) = \sigma_{\max}\left(\frac{x}{d_{\text{PML}}}\right)^m, \tag{22}$$

where d_{PML} is the thickness of the PML, $\sigma_{\max}$ is the maximum conductivity, and m is the grading order. By carefully selecting these parameters, the PML can effectively absorb outgoing waves over a broad range of angles and frequencies, ensuring minimal reflection and accurate simulation results.

1.4 Gaussian Sinusoidal Pulse

In FDTD simulations, a Gaussian sinusoidal pulse is often used as a source due to its localized and broadband nature. The pulse is defined by a Gaussian envelope modulating a sinusoidal carrier wave, which can be mathematically represented as [4]:

$$E(t) = E_0 \exp\left(-\frac{(t - t_0)^2}{2\sigma^2}\right) \sin(2\pi f_c t), \tag{23}$$

where E_0 is the peak amplitude, t_0 is the time delay, σ is the pulse width, and f_c is the central frequency of the sinusoidal wave. The Gaussian envelope ensures that the pulse is localized in time, while the sinusoidal component determines its frequency content.

In the context of FDTD, the source can be introduced into the simulation domain by updating the electric field component at a specific grid point. This allows the study of the pulse's interaction with various media and boundaries.

This implies that changes in energy density within a volume are due to the net flux of energy across the volume's boundaries. The stabilization of energy in the graph indicates conservation of internal energy after accounting for boundary absorption [2].

In this work, a circular dielectric medium is introduced into the 2D FDTD computational domain to investigate the interaction of electromagnetic waves with a dielectric object as illustrated in Fig. 1. The dielectric cylinder is characterized by a relative permittivity (ϵ_r) of 30 and a conductivity (σ) of $0.3\,\text{S}\,\text{m}^{-1}$. The diameter of the cylinder is 20 cm, representing both lossless and lossy linear isotropic homogeneous media (LIHMs).

A *lossless* linear isotropic homogeneous medium (LIHM) is one where the material has zero conductivity ($\sigma = 0\,\text{S}\,\text{m}^{-1}$). In such a medium, electromagnetic waves can propagate without any energy loss. The material's properties, like relative permittivity (ϵ_r), affect the speed and direction of the wave, but not its amplitude.

In contrast, a *lossy* LIHM is a material with non-zero conductivity ($\sigma > 0\,\text{S}\,\text{m}^{-1}$). In this case, the medium absorbs some of the electromagnetic wave's energy, leading to attenuation of the wave as it propagates. The dielectric cylinder in this work, characterized by $\sigma = 0.3\,\text{S}\,\text{m}^{-1}$, is an example of a lossy medium, which allows the study of both idealized (lossless) and realistic (lossy) conditions.

The presence of the dielectric cylinder affects the propagation of the electromagnetic waves, causing phenomena such as reflection, refraction, and scattering. These effects are crucial for understanding the behavior of waves in real-world applications, such as radar and wireless communications.

The update equations for the FDTD simulation in the presence of a dielectric medium are modified to account for the material properties. For the electric field component E_z, the update equation reads

$$E_z^{n+1}(i,j) = \frac{1-\frac{\sigma\Delta t}{2\epsilon}}{1+\frac{\sigma\Delta t}{2\epsilon}} E_z^n(i,j) + \frac{\Delta t}{\epsilon\left(1+\frac{\sigma\Delta t}{2\epsilon}\right)}\left[\frac{H_y^n(i,j)-H_y^n(i-1,j)}{\Delta x} - \frac{H_x^n(i,j)-H_x^n(i,j-1)}{\Delta y}\right], \tag{24}$$

where ϵ is the permittivity of the dielectric medium, Δt is the time step, and Δx and Δy are the spatial steps in the x and y directions, respectively.

For the magnetic field components H_x and H_y, the update equations remain the same as in the free space case, as the magnetic permeability is assumed to be that of free space ($\mu = \mu_0$)

$$H_x^{n+1/2}(i,j) = H_x^{n-1/2}(i,j) - \frac{\Delta t}{\mu_0}\frac{E_z^n(i,j+1)-E_z^n(i,j)}{\Delta y}, \tag{25}$$

$$H_y^{n+1/2}(i,j) = H_y^{n-1/2}(i,j) + \frac{\Delta t}{\mu_0}\frac{E_z^n(i+1,j)-E_z^n(i,j)}{\Delta x}. \tag{26}$$

By integrating the circular dielectric medium into the FDTD simulation, the study aims to provide insights into the complex interactions of electromagnetic waves with dielectric materials. Understanding these interactions is essential for the design and analysis of various electromagnetic systems, such as radar and wireless communication systems. The simulation helps to visualize and quantify the effects of the dielectric medium on wave propagation, enabling better prediction and optimization of system performance [4].

The algorithm for simulations of electromagnetic wave propagation in dielectric media is given in the Algorithm Box 1.

Algorithm 1: 2D FDTD Electromagnetic Wave Simulation with PML and Dielectric Medium

Data: Simulation parameters: dx, dy, dt, Lx, Ly, T
Result: Animation and plots of electromagnetic wave propagation
Initialization:
– Define constants: c_0, ϵ_0, μ_0
– Compute grid size Nx, Ny, time steps Nt
– Initialize permittivity matrix ϵ_r and conductivity matrix σ
– Set up circular dielectric medium in ϵ_r and σ
– Initialize field arrays E_z, H_x, H_y
– Define source parameters: position, pulse width, frequency f_0
– Apply PML boundary conditions to σ
Main Simulation Steps:
for *each time step n from* 0 *to* $Nt - 1$ **do**
    – Calculate current time $t = n \times dt$ for each time step n
    – Compute source pulse for time t
end
– Update the electric field E_z at the source position using:

$$E_z^{n+1}(i,j) = \frac{1 - \frac{\sigma \Delta t}{2\epsilon_0}}{1 + \frac{\sigma \Delta t}{2\epsilon_0}} E_z^n(i,j) + \frac{\Delta t}{\epsilon \left(1 + \frac{\sigma \Delta t}{2\epsilon_0}\right)} \times \left[\frac{H_y^n(i,j) - H_y^n(i-1,j)}{\Delta x} - \frac{H_x^n(i,j) - H_x^n(i,j-1)}{\Delta y} \right]$$

– Update magnetic fields H_x, H_y using:

$$H_x^{n+1/2}(i,j) = H_x^{n-1/2}(i,j) - \frac{\Delta t}{\mu_0} \left(\frac{E_z^n(i,j+1) - E_z^n(i,j)}{\Delta y} \right)$$

$$H_y^{n+1/2}(i,j) = H_y^{n-1/2}(i,j) + \frac{\Delta t}{\mu_0} \left(\frac{E_z^n(i+1,j) - E_z^n(i,j)}{\Delta x} \right)$$

– Apply PML conditions and update fields using:

$$D_z^{n+1/2}(i,j) = g_{i3}(i) g_{j3}(j) D_z^{n-1/2}(i,j) + 0.5 g_{i2}(i) g_{j2}(j) \times \left[H_y^n\left(i + \frac{1}{2}, j\right) - H_y^n\left(i - \frac{1}{2}, j\right) - H_x^n\left(i, j + \frac{1}{2}\right) + H_x^n\left(i, j - \frac{1}{2}\right) \right].$$

– Store results every 10 steps and calculate total energy: **if** *n mod 10 == 0* **then**
    – Store current E_z field for animation
    – Compute and store total energy in the system
end
Visualization:
– Create 2D animation of E_z over time
– Generate 3D surface plots of E_z at selected time steps
– Plot total electromagnetic energy as a function of time

1.5 Problem Description

The work is based on simulating 2D FDTD electromagnetic wave propagation in a dielectric medium. The simulation will incorporate advanced features and visualizations of reflection, refraction, and absorption. The simulation setup includes various physical constants, simulation parameters, material properties, field initialization, source configuration, and PML boundary conditions (BC). The parameters chosen are given in Table 1.

In this simulation, vacuum space is considered alongside a dielectric medium to study the behavior of electromagnetic waves under ideal conditions. The inclusion of vacuum allows us to observe the wave's natural behavior, minimizing external influences such as dispersion and loss. This provides a clear baseline for understanding wave phenomena such as reflection and refraction.

Table 1 Summary of constants, material properties, source configuration, and PML BCs

Parameters	Values/Types	Units
Constants		
Speed of light (c_0)	3×10^8	$\mathrm{m\,s^{-1}}$
Vacuum permittivity (ϵ_0)	8.54×10^{-12}	$\mathrm{F\,m^{-1}}$
Vacuum permeability (μ_0)	$4\pi \times 10^{-7}$	$\mathrm{H\,m^{-1}}$
Material properties		
Relative permittivity (ϵ_r)	30	–
Conductivity (σ)	0.30	$\mathrm{S\,m^{-1}}$
Diameter of dielectric cylinder	0.20	m
Simulation parameters		
Spatial step in x (Δx)	0.01	m
Spatial step in y (Δy)	0.01	m
Time step (Δt)	$0.9 \times \dfrac{\Delta x}{c_0\sqrt{2}}$	s
Domain length in x (L_x)	2.0	m
Domain length in y (L_y)	2.0	m
Position of dielectric center	1.0, 1.0	m
Total simulation time (T)	1×10^{-7}	s
Source configuration		
Position	$\left(\dfrac{N_x}{4}, \dfrac{N_y}{2}\right) = (0.5, 1.0)$	m
Pulse width	10^{-10}	s
Frequency (f_0)	5×10^6	Hz
PML BC		
Thickness	20	cells
Max conductivity ($\sigma_{\max}$)	0.1	$\mathrm{S\,m^{-1}}$

The dielectric medium, characterized by its relative permittivity and conductivity, is included within the same simulation environment to observe how these material properties affect wave propagation. By studying the interaction of waves with the dielectric medium, we can analyze how it alters wave behavior in terms of reflection, refraction, and absorption. This comprehensive approach ensures that the effects of the dielectric medium on wave behavior are clearly understood within a controlled simulation environment.

2 Prompt

To effectively utilize ChatGPT for formulating and solving a 2D FDTD electromagnetic wave propagation simulation in a dielectric medium, the prompt must be structured with specific instructions and provide comprehensive and precise inputs. The goal is to ensure that the problem is clearly articulated and all necessary parameters are included to achieve a precise and correct solution. The task involves defining the physical constants, simulation parameters, material properties, field initialization, source configuration, PML BC, and analysis requirements perfectly. The prompt must specify the following details

- **Physical Constants**: Specify the speed of light, vacuum permittivity, and vacuum permeability.
- **Simulation Parameters**: Describe the spatial steps, time step, domain size, and simulation time.
- **Material Properties**: Define the relative permittivity and conductivity of the dielectric medium, as well as its geometric placement in the domain.
- **Field Initialization**: Specify the initialization of the electric and magnetic fields.
- **Source Configuration**: Describe the source type ,position, pulse width, and frequency.
- **PML BC**: Mention the thickness and maximum conductivity of the PML boundaries.
- **Units**: Ensure all inputs are given in consistent units to avoid errors in the simulation.

The prompt must also include a set of instructions or guidelines to ensure that ChatGPT understands the task clearly and displays the results as required. The following instructions are provided for this particular field of study:

- **FDTD Setup**: Instructions to define the spatial and time steps, initialize the field arrays, and apply the FDTD update equations for both electric and magnetic fields. Instructions to incorporate the material properties and PML BC into the simulation.
- **Source Implementation**: Instructions to define the source configuration, including its position, pulse width, and frequency.
- **Post-Processing**: Instructions to compute the outputs like electric and magnetic field distributions, visualize the fields over time, and analyze the wave interactions

with the dielectric medium. Wherever applicable, formulas or pseudo-code can be provided to minimize the assumption of ChatGPT.

- **Visualization**: The format of the output is mentioned in the prompt. Instructions provided to generate plots or animations of the field distributions with proper legends, labels, and color maps for better understanding.
- **Instruction to ask ChatGPT for comments in the code**: Request comments within the code to explain the steps and calculations performed.

As a computational physicist, generate a complete Python program for a 2D Finite-Difference Time-Domain (FDTD) electromagnetic wave propagation in dielectric medium simulation. The program must include the following components and features, with the same function names as mentioned in this script.

1. **Import necessary libraries**:
   - `numpy` for numerical computations,
   - `matplotlib` for plotting and animation,
   - `cm` from `matplotlib`.
2. **Define physical constants**:
   - Speed of light in vacuum ($c_0 = 3 \times 10^8\ \mathrm{m\,s^{-1}}$),
   - Permittivity of free space ($\epsilon_0 = 8.54 \times 10^{-12}\ \mathrm{F\,m^{-1}}$),
   - Permeability of free space ($\mu_0 = 4\pi \times 10^{-7}\ \mathrm{H\,m^{-1}}$).
3. **Set up simulation parameters**:
   - Spatial steps ($\Delta x, \Delta y$) = (0.01, 0.01 m),
   - Time step (Δt) calculated for stability = $0.9 \times \frac{\Delta x}{c_0\sqrt{2}}$,
   - Domain dimensions (L_x, L_y) = (2, 2 m),
   - Total simulation time (T) = 1×10^{-7} s,
   - Calculate grid sizes (N_x, N_y) and number of time steps (N_t).
4. **Create material properties**:
   - Define a circular dielectric medium in the center of the domain: ϵ_r, σ,
   - Center: $center\_x = 0.5 \times N_x$, $center\_y = 0.5 \times N_y$,
   - Radius: $radius = \frac{0.20}{\Delta x}$,
   - Set relative permittivity (ϵ_r) = 30 and conductivity (σ) = 0.30 $\mathrm{S\,m^{-1}}$ for the medium.
5. **Initialize field arrays**:
   - Electric field (E_z),
   - Magnetic fields (H_x, H_y).

6. **Set up source parameters**:
   - Source position: ($N_x//4$, $N_y//2$),
   - Pulse width: 1×10^{-10} s,
   - Frequency: 5×10^6 Hz.
7. **Set up PML Boundary Conditions (BC)**:
   - PML thickness: 20 cells,
   - Maximum conductivity ($\sigma_{\max}$): 0.1 $\mathrm{S\,m^{-1}}$.

   Create a function `apply_pml()` using a `for` loop in the range of `pml_thickness`. This function applies PML BC to the conductivity array `sigma` in a 2D grid, which is used in FDTD simulations to absorb outgoing waves at the boundaries and prevent reflections.
8. **Create update functions for the electromagnetic fields**:
   - Includes updating the magnetic fields H_x and H_y, updating the electric field E_z in a single function, then applying conductivity for absorption.
9. **Simulation functions**:
   - Create a function `run_simulation()`: This function runs a simulation of electromagnetic wave propagation and captures snapshots of the electric field at regular intervals. The function should propagate a Gaussian-modulated sinusoidal pulse wave in a medium characterized by its relative permittivity (ϵ_r) and electrical conductivity (σ). It captures the electric field (E_z) at every 10th time step and stores these snapshots as 'frames' in a list for subsequent analysis or visualization.
   - Create a function `animate(frames, title)`: Create an animation of the electric field from a series of frames and save it as a GIF or MP4 file.
   - Create a function `run_simulation_3D_plot()`: Capture snapshots of the electric field (E_z) at specific time frames(100,500,1000,1200) for 3D plotting.
   - Create a function `plot_3d_surface(Ez, title)`: Plot a 3D surface of the electric field component E_z.
   - Create a function `calculate_energy(Ez, Hx, Hy, eps_r, dx, dy)`: Calculate the total electromagnetic energy in the system.
   - Create a function `run_simulation_with_energy()`: Run a simulation and track the energy of the system at regular intervals.
10. **Run the simulation and plot the results**:
    - Call the function `energies = run_simulation_with_energy()` and plot the results with appropriate labels and titles.

- Call the function `frames = run_simulation()`.
- Call the function `frames_1 = run_simulation_3D_plot()`.
- Call the function `animate(frames, '2D FDTD Simulation of Gaussian Pulse')`.
- Plot the 3D surface plots for specific time intervals.

3 Code Listing

```

#instruction to run the code.
#* line 104 and 106 , allows user to save GIF and animated video of this visualization.
#* it can be best visualized if you run line 104 as GIF ,as Video processing time may take a while.



import numpy as np
import matplotlib.pyplot as plt
import matplotlib.animation as animation
from matplotlib import cm
from mpl_toolkits.mplot3d import Axes3D

# Constants
c0 = 3e8  # Speed of light in vacuum
eps0 = 8.854e-12  # Permittivity of free space
mu0 = 4 * np.pi * 1e-7  # Permeability of free space

# Simulation parameters
dx = 0.01  # Spatial step (m)
dy = 0.01  # Spatial step (m)
dt = 0.9 * dx / (c0 * np.sqrt(2))  # Time step (s)
Lx = 2.0  # Domain length in x (m)
Ly = 2.0  # Domain length in y (m)
T = 1e-7  # Total simulation time (s)

Nx = int(Lx / dx)  # Number of spatial steps in x
Ny = int(Ly / dy)  # Number of spatial steps in y
Nt = int(T / dt)  # Number of time steps

# Material properties for circular dielectric medium
eps_r = np.ones((Nx, Ny))
sigma = np.zeros((Nx, Ny))
center_x = int(0.5* Nx)
center_y = int(0.5 * Ny)
```

```
radius = int(0.20 / dx)

for i in range(Nx):
    for j in range(Ny):
        if (i - center_x)**2 + (j - center_y)**2 <= radius**2:
            eps_r[i, j] = 30
            sigma[i, j] = 0.3

# Field arrays
Ez = np.zeros((Nx, Ny))
Hx = np.zeros((Nx, Ny))
Hy = np.zeros((Nx, Ny))

# Source parameters
source_position = (Nx // 4, Ny // 2)
pulse_width = 1e-10

f0 = 5e6 # Frequency (Hz)

# PML boundary conditions
pml_thickness = 20  # Number of cells in PML
sigma_max = 0.1  # Maximum conductivity in PML

def apply_pml(sigma, pml_thickness, sigma_max):
    for i in range(pml_thickness):
        sigma[i, :] = sigma_max * (pml_thickness - i) /
    ↪ pml_thickness
        sigma[-i-1, :] = sigma_max * (pml_thickness - i) /
    ↪ pml_thickness
        sigma[:, i] = sigma_max * (pml_thickness - i) /
    ↪ pml_thickness
        sigma[:, -i-1] = sigma_max * (pml_thickness - i) /
    ↪ pml_thickness

apply_pml(sigma, pml_thickness, sigma_max)

def update_fields(Ez, Hx, Hy, eps_r, sigma):
    # Update magnetic fields Hx and Hy
    Hx[:, :-1] -= (dt / mu0 / dy) * (Ez[:, 1:] - Ez[:, :-1])
    Hy[:-1, :] += (dt / mu0 / dx) * (Ez[1:, :] - Ez[:-1, :])

    # Update electric field Ez
    Ez[1:-1, 1:-1] += (dt / eps0 / eps_r[1:-1, 1:-1]) * (
        (Hy[1:-1, 1:-1] - Hy[:-2, 1:-1]) / dy -
        (Hx[1:-1, 1:-1] - Hx[1:-1, :-2]) / dx
    )

    # Apply conductivity for absorption
    Ez *= np.exp(-sigma * dt / eps0)

def run_simulation():
    frames = []
    for n in range(Nt):
```

```
        t = n * dt
        pulse = np.exp(-((t - 4 * pulse_width) ** 2) / (
    ↪ pulse_width ** 2)) * np.cos(2 * np.pi * f0 * t)
        Ez[source_position] += pulse

        update_fields(Ez, Hx, Hy, eps_r, sigma)

        if n % 10 == 0:
            frames.append(np.copy(Ez))

    return frames
def animate(frames, title):
    fig, ax = plt.subplots()

    def update(frame):
        ax.clear()
        im = ax.imshow(frame.T, cmap=cm.viridis, vmin=-0.01, vmax
    ↪ =0.01, animated=True)
        ax.set_title(title)
        return [im]

    ani = animation.FuncAnimation(fig, update, frames=frames,
    ↪ interval=50, blit=True)
    #Save as GIF using Pillow
    #ani.save("2D_FDTD_Simulation.gif", writer='pillow')
    #Save as MP4 video using ffmpeg
    #ani.save("2D_FDTD_Simulation.mp4", writer='ffmpeg')
    plt.show()
    return ani

def run_simulation_3D_plot():
    frames_1 = []
    time_frames = [ 100, 500,1000,1200]  # Time frames for
    ↪ capturing the results
    for n in range(Nt):
        t = n * dt
        pulse = np.exp(-((t - 4 * pulse_width) ** 2) / (
    ↪ pulse_width ** 2)) * np.cos(2 * np.pi * f0 * t)
        Ez[source_position] += pulse

        update_fields(Ez, Hx, Hy, eps_r, sigma)

        if n in time_frames:
            frames_1.append(np.copy(Ez))

    return frames_1
def plot_3d_surface(Ez, title):
    fig = plt.figure()
    ax = fig.add_subplot(111, projection='3d')
    X, Y = np.meshgrid(np.arange(Ez.shape[0]), np.arange(Ez.shape
    ↪ [1]))
    surf = ax.plot_surface(X, Y, Ez.T, cmap=cm.viridis,edgecolor=
    ↪ 'black')
```

```
    ax.set_title(title)
    ax.set_xlabel('X')
    ax.set_ylabel('Y')
    ax.set_zlabel('Electric Field (Vm$^{-1}$)')

    plt.show()

def calculate_energy(Ez, Hx, Hy, eps_r, dx, dy):
    electric_energy = 0.5 * eps0 * np.sum(eps_r * Ez**2) * dx * dy
    magnetic_energy = 0.5 * np.sum((Hx**2 + Hy**2) / mu0) * dx * dy
    total_energy = electric_energy + magnetic_energy
    return total_energy

def run_simulation_with_energy():
    energies = []
    for n in range(Nt):
        t = n * dt
        pulse = np.exp(-((t - 4 * pulse_width) ** 2) / (pulse_width ** 2)) * np.cos(2 * np.pi * f0 * t)
        Ez[source_position] += pulse

        update_fields(Ez, Hx, Hy, eps_r, sigma)

        if n % 10 == 0:
            energy = calculate_energy(Ez, Hx, Hy, eps_r, dx, dy)
            energies.append(energy)
    return energies
# Run the simulations and plotting
energies = run_simulation_with_energy()
plt.figure()
plt.plot(np.arange(0, Nt, 10) * dt, energies)
plt.xlabel('Time (s)')
plt.ylabel('Total Electromagnetic Energy (J)')
plt.title('Energy Conservation in 2D FDTD Simulation')
plt.grid(True)
plt.show()
# Run simulation and create animation , create 3D surface plots
frames = run_simulation()
frames_1 = run_simulation_3D_plot()

time_labels = [100 , 500 , 1000 , 1200]
animate(frames, '2D FDTD Simulation of Gaussian Pulse')
for frame_1, label in zip(frames_1, time_labels):
    plot_3d_surface(frame_1, f'2D FDTD Simulation at T={label:} s')
```

Listing 1 Output by ChatGPT for solving Electromagnetic wave propagation in dielectric medium

4 Tests for Verification

The most important part of the work is to verify the program generated by ChatGPT-4 and ensure that the results conform with the solution of the problem stated in the problem description.

This work's verification is done by comparing the FDTD simulation results with those presented in other technical papers and through self-verification. The FDTD program generated by ChatGPT is run on an Integrated Development Environment (IDE). It generates visualizations of the electromagnetic wave propagation, including reflection, refraction, and absorption, which are compared to the expected results. The comparison involves examining the accuracy of the field distributions and energy conservation over time.

Next, I will discuss the verification done by matching similar graphs from research papers. The results from the FDTD simulation are compared to those from established research to ensure the accuracy and validity of the ChatGPT-generated code.

4.1 Verification: Graphically

The result from the FDTD simulation at time frame $t = 1200$ s aligns well with the outcomes observed in similar studies, as discussed in [5]. The symmetry, wave interaction pattern, and boundary conditions all reflect the accuracy of the simulation generated by the FDTD program.

The Image from [5] shows a simulation of a plane wave impinging on a dielectric cylinder at time frame $t = 75$ s, while Fig. 5 from the FDTD simulation of a Gaussian wave at time frame $t = 1200$ s. Despite different waveforms and simulation parameters, both results demonstrate consistent interactions with the dielectric media, enabling a comparative analysis of wave reflection and refraction patterns.

- **Wave Interaction Pattern**: Both visualizations show two distinct peaks with a valley in between, indicating similar wave interaction patterns with the medium.
- **Symmetry**: The wave propagation appears symmetric in both images, suggesting correct modeling of the medium's properties and boundaries.
- **Boundary Conditions**: Both simulations use absorbing boundary conditions (PML), as evident from the absence of reflected waves at the edges.

These similarities validate the FDTD simulation approach employed and confirm its reliability in modeling electromagnetic wave interactions in dielectric media.

4.1.1 Verification: Energy Conservation Analysis

The prompt has its own way of validating the program it has generated with the method of Energy Conservation as discussed below.

The graph in Fig. 6 shows the total electromagnetic energy as a function of time. The corresponding snapshots in Fig. 7 visually depict the wave's interaction with the dielectric medium at different time steps. Key observations include:

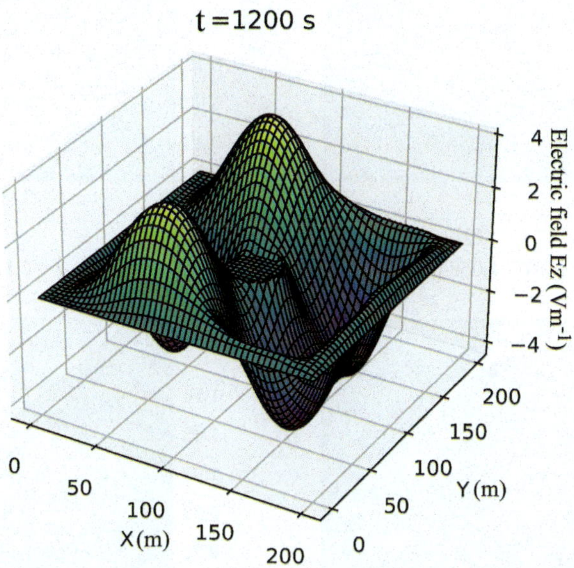

Fig. 5 3D visualization of the electric field component E_z generated using a 2D Finite-Difference Time-Domain (FDTD) simulation at time frame $t = 1200$ s. The plot illustrates the propagation of the electromagnetic wave within a dielectric medium, showcasing the complex interactions of the wave as it reflects and refracts within the medium

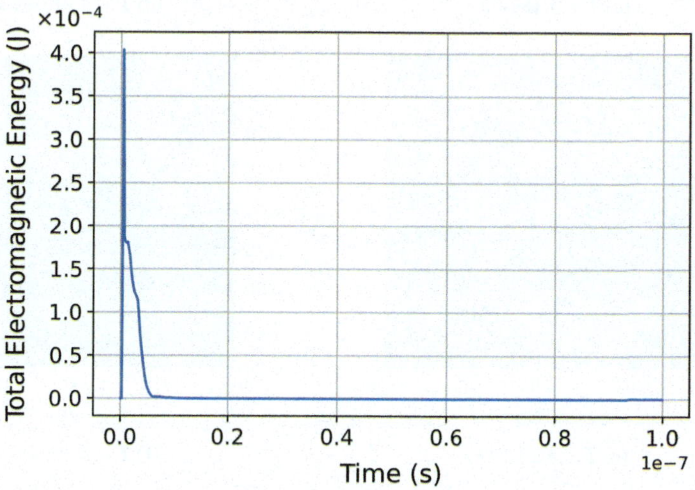

Fig. 6 Total electromagnetic energy as a function of time in a 2D FDTD simulation. The plot demonstrates energy conservation within the computational domain and validates the accuracy of the FDTD simulation over time

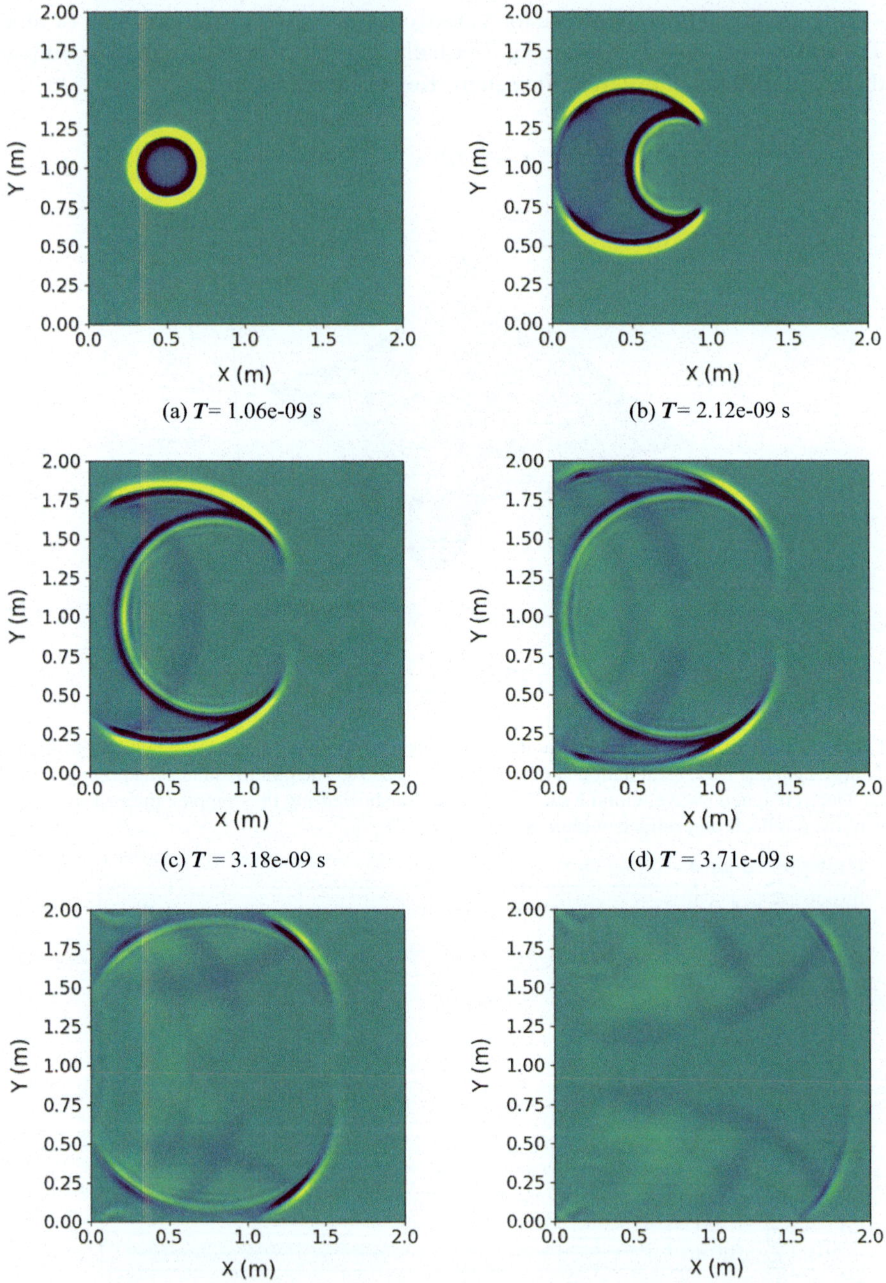

(a) T= 1.06e-09 s (b) T= 2.12e-09 s

(c) T = 3.18e-09 s (d) T = 3.71e-09 s

(e) T = 4.24e-09 s (f) T = 5.30e-09 s

Fig. 7 These time-lapse visualizations illustrate the reflection, refraction, and absorption of the electromagnetic wave as it interacts with the dielectric medium and exits the simulation domain. The progression from the initial pulse propagation to wave dissipation provides insights into energy conservation and wave dynamics within the medium

- **Initial Energy Peak**: The energy starts at a peak value due to the initial source input, as seen in Fig. 7a, where the wave is centered within the dielectric medium.
- **Energy Decay**: Rapid decay in the initial phase, likely due to absorption by the PML boundaries, can be correlated with Fig. 7c and d, where the wave begins to exit the medium and is partially absorbed by the boundaries.
- **Energy Conservation**: The energy drops down to zero after the initial decay, indicating the wave has left the simulation box, such that a steady state is restored where internal energy is conserved. This is visually supported by Fig. 7f, where the wave has exited the simulation domain, leaving behind minimal residual energy.

According to the Poynting theorem, the energy density u and the Poynting vector $\boldsymbol{S}$ satisfy

$$\frac{\partial u}{\partial t} + \nabla \cdot \boldsymbol{S} = -\boldsymbol{J} \cdot \boldsymbol{E}. \tag{27}$$

For a lossless medium with no free currents ($\boldsymbol{J} = 0$), this simplifies to

$$\frac{\partial u}{\partial t} + \nabla \cdot \boldsymbol{S} = 0. \tag{28}$$

5 Discussion

After the ChatGPT-4 generating the program, there are several aspects to verify before running it in any Integrated Development Environment (IDE). These aspects are summarized as follows:

- **Completeness of the program**: You must confirm that ChatGPT 4 has supplied the entire program for this kind of work, where you need to demonstrate outcomes rather than generate numbers. The main problem is that each time you enter the prompt, a slightly different version of the prompt is generated, making it difficult to anticipate the outcome.
- **In Steps**: Occasionally, ChatGPT-4 produces code in a step-by-step fashion instead than as a single, integrated application. If this happens, request the whole program.
- **Short term memory capacity**: There is a limit to ChatGPT-4's output where it outputs precisely what the prompt instructs; after that, you must prompt again to concentrate on the details provided in the first prompt. Claiming that it is accurate, the most recent version Chatgpt-4o can read the prompt in its entirety and produce a program that is as exact as possible; based on the prompt, it can provide the desired output based on an accurate user prompt.
- **Updating and frames**: As a visualization objective, it its mandatory to check that the updated electric field equation captures the frames at every step for a smooth simulation.

- **Learned Lessons**: Chatgpt-4 is advanced for solving complex tasks compared to any other model that exists. The more references and resources you provide, the better your accuracy. Comparing the two models Chatgpt-4 and 4o, the newer model delivers precisely and more than what you expect; it can handle a large amount of data compared to Chatgpt-4. If you expect complex mathematical reasoning, version 4 is worth considering.

6 Conclusion

The work aimed to generate a Python program using ChatGPT-4 for simulating electromagnetic wave propagation in a dielectric medium using the 2D Finite-Difference Time-Domain (FDTD) method. The objective was to examine wave interactions such as reflection, transmission, and absorption, within different dielectric environments.

It was observed that ChatGPT-4 requires comprehensive and precise inputs to perform accurately, as discussed in the prompt section. Detailed task descriptions and explicit instructions are necessary to ensure that ChatGPT-4 does not make any assumptions about key factors in the FDTD simulation setup.

Comparing the FDTD program generated by ChatGPT-4 with theoretical models and established methods that the solutions were correctly formulated. The results of the simulations, including the electric and magnetic field distributions, were consistent with theoretical predictions and comparable to those produced by other established ways.

Verification with graph solutions indicated that the FDTD results converged with increased grid resolution. The simulation of the Gaussian sinusoidal pulse and its interaction with a cylindrical dielectric medium provided accurate visualizations of wave propagation phenomena. The implementation of PML BC effectively minimized reflections at the boundaries, further validating the simulation's accuracy.

The energy conservation analysis confirmed the physical accuracy of the generated code, with the total energy stabilizing over time after initial absorption by the PML boundaries. The overall performance of the ChatGPT-4 generated FDTD simulation demonstrated high accuracy and reliability.

In summary, this study validates ChatGPT-4's ability to produce FDTD simulation programs that are accurate and dependable, greatly simplifying and improving the process of resolving challenging electromagnetic wave propagation issues. This effective application of AI in computational electromagnetics shows how broadly applicable it may be in scientific computing and provides a viable method for automating the creation of intricate simulation tools.

References

1. D. Sarkar, "FDTD Analysis of Guided Electromagnetic Wave Interaction with Time-Modulated Dielectric Medium," *Springer Nature Singapore, 2022*. ISBN: 978-981-19-1630-4, https://doi.org/10.1007/978-981-19-1630-4.
2. J. Chen and Y. Kougong, "Energy and momentum of electromagnetic waves in media," *Journal of Optics*, 2023, https://doi.org/10.1007/s12596-023-01183-0.
3. N. Faruk and U. M. Gana, "FDTD Modelling of Electromagnetic waves in Stratified Medium," *Global Journal of Engineering Research*, 2013, https://doi.org/10.4314/gjer.v12i1.1.
4. D. M. Sullivan, "Electromagnetic Simulation Using the FDTD Method," *Wiley-IEEE Press, 2013*. ISBN: 9781118646700, https://doi.org/10.1002/9781118646700.
5. E. Balti, "Finite-Difference Time-Domain Simulations," *engrxiv*, 2021, https://doi.org/10.31224/osf.io/yjcs8.
6. K. Yee, "Numerical solution of initial boundary value problems involving Maxwell's equations in isotropic media," *IEEE Transactions on Antennas and Propagation*, vol. 14, no. 3, pp. 302–307, 1966, https://doi.org/10.1109/TAP.1966.1138693.
7. J. X. Liu, Z. K. Yang, L. Ju, et al. "Boltzmann Finite-Difference Time-Domain Method Research Electromagnetic Wave Oblique Incidence into Plasma." *Plasmonics 13, 1699–1704*, 2018, https://doi.org/10.1007/s11468-017-0681-3.

References

Flow Around an Obstacle Using the Lattice Boltzmann Method

Fenil Lathiya and Bernhard Eidel

Abstract This chapter delves into the generation of Python code by ChatGPT-4o for 2D simulations of fluid flow around obstacles of circular or square shape. The numerical solution is achieved using the Lattice Boltzmann Method (LBM) with the D2Q9 model. A significant portion of the chapter is dedicated to the intricacies of prompt engineering for this specific task, as well as the thorough verification of the generated code. The performance of the code is rigorously tested against results obtained from a commercial solver, focusing on velocity fields, pressure fields, and pressure distribution. The remarkable quantitative agreement across all criteria demonstrates the effectiveness of GPT-4o in producing accurate and reliable code for complex fluid dynamics simulations.

1 Introduction

The study of fluid dynamics involves understanding the behavior of fluid flow around obstacles, a task that requires advanced computational methods. The Lattice Boltzmann Method (LBM) is such a method in Computational Fluid Dynamics (CFD) through a lattice grid-based approach. This method models fluid behavior at a microscopic level using particle distribution functions and kinetic theory, which can then be translated into macroscopic fluid properties. For a comprehensive overview of the Lattice Boltzmann Method and its applications, we refer to [1] and for a sound introduction to the LBM in fluid flows [2].

This paper is organized into several sections to comprehensively discuss the AI-Assisted Coding Project. After this introduction, the next section details the Formulation of the Coding Task, explaining the problem setup and the approach taken.

F. Lathiya · B. Eidel (✉)
M5, IMFD, TUBAF, Freiberg, Germany
e-mail: bernhard.eidel@imfd.tu-freiberg.de

F. Lathiya
e-mail: Fenil-Maganbhai.Lathiya@student.tu-freiberg.de

© The Author(s), under exclusive license to Springer Nature Switzerland AG 2025
B. Eidel (ed.), *GPT for Python-Coding in Computational Materials Science and Mechanics*, Studies in Computational Intelligence 1198,
https://doi.org/10.1007/978-3-031-85470-5_10

Following this, the Code Listing section presents the AI-generated Python code for the LBM simulation. The Code Testing/Performance section evaluates the performance and accuracy of the code through tests and comparisons with ANSYS Fluent. The Discussion section provides an analysis of the results and insights gained from the project. Finally, the Conclusions section summarizes the findings and implications of the study. Additional details and data are included in the Appendix.

1.1 Lattice Boltzmann Method: Theory and Modeling

The main idea behind the LBM is to represent the fluid using particles that move and collide each other on a discrete lattice grid. The key part of this method lies in the Boltzmann equation, which describes the evolution of the particle distribution function. A critical component of this equation is the collision step [7], which is given by

$$f_{\text{out}} = f_{\text{in}} - \omega \left(f_{\text{in}} - f_{\text{eq}} \right), \tag{1}$$

where f_{in} is the incoming particle distribution function, f_{out} is the outgoing particle distribution function after the collision, ω the relaxation parameter, and f_{eq} the equilibrium distribution function.

LBM is advantageous due to its simplicity in handling complex boundary conditions (BC) and its flexibility in simulating various types of flows, which makes it ideal for studying fluid flow around obstacles. The method translates the microscopic particle dynamics into macroscopic flow properties, such as velocity and pressure fields, by averaging them.

In LBM, the fluid is modeled as a collection of particles that reside on the nodes of a discrete lattice. These particles propagate to neighboring nodes and collide, redistributing their velocities according to predefined rules. The macroscopic properties of the fluid, such as density and velocity, are obtained by taking moments of the particle distribution functions.

One of the significant advantages of LBM is its ability to handle complex geometries and BC with ease. Traditional CFD methods often require sophisticated meshing techniques and BC formulations. In contrast, LBM uses simple bounce-back rules to impose no-slip BC on solid surfaces, making it more straightforward to implement for problems involving obstacles.

The standard lattice model used in LBM is the D2Q9 model (see Sect. 1.2) for two-dimensional flows. In this model, each node in the lattice has nine possible velocity vectors, including one stationary and eight moving in different directions. The particle distribution function at each node evolves according to the LBM equation, [3]

$$f_i(\boldsymbol{x} + \boldsymbol{e}_i, t + 1) = f_i(\boldsymbol{x}, t) + \Omega_i, \tag{2}$$

where f_i is the particle distribution function in the direction $\boldsymbol{e}_i$, and Ω_i represents the collision operator that models the redistribution of particles due to collisions. The collision operator Ω_i is often represented by the Bhatnagar-Gross-Krook (BGK) approximation, which simplifies to [7]:

$$\Omega_i = -\frac{1}{\tau}\left(f_i(\mathbf{x}, t) - f_i^{\mathrm{eq}}(\mathbf{x}, t)\right). \tag{3}$$

Here, τ is the relaxation time parameter, and f_i^{eq} is the local equilibrium distribution function. The BGK approximation ensures that the system relaxes towards equilibrium over time, with the rate of relaxation governed by τ. The collision operator Ω_i simplifies the collision term to a relaxation towards a local equilibrium distribution f_i^{eq}. This relaxation process occurs over a characteristic time scale τ, ensuring that the system gradually approaches equilibrium.

The equilibrium distribution function f_{eq} for the D2Q9 model is given by [4]

$$f_{\mathrm{eq},i} = w_i \rho \left[1 + \frac{\boldsymbol{e}_i \cdot \boldsymbol{u}}{c_s^2} + \frac{(\boldsymbol{e}_i \cdot \boldsymbol{u})^2}{2c_s^4} - \frac{u^2}{2c_s^2}\right], \tag{4}$$

where w_i are the weights associated with each direction, ρ is the fluid density, $\boldsymbol{u}$ is the macroscopic velocity, and c_s is the speed of sound in the lattice.

The macroscopic fluid properties, such as density ρ and velocity $\boldsymbol{u}$ are obtained by taking moments of the particle distribution function [4]

$$\rho = \sum_i f_i, \tag{5}$$

$$\rho \boldsymbol{u} = \sum_i f_i \boldsymbol{e}_i. \tag{6}$$

1.2 The D2Q9 Model

The D2Q9 model, an abbreviation for "Two-Dimensional, Nine-Directional", is a prevalent lattice model employed within the LBM for simulating fluid flows in two dimensions. This model is crucial in CFD for its efficiency in simulating complex fluid behaviors at the macroscopic level.

Structure and velocity vectors: The D2Q9 model is constructed around a square lattice, where each node is linked to its nearest and next-nearest neighbors, enabling a comprehensive representation of fluid flow directions. The connectivity and possible movement directions at each node are depicted in Fig. 1:

- $\boldsymbol{e}_0 = (0, 0)$: Represents the rest particle.
- $\boldsymbol{e}_1 = (1, 0)$, $\boldsymbol{e}_2 = (0, 1)$, $\boldsymbol{e}_3 = (-1, 0)$, $\boldsymbol{e}_4 = (0, -1)$: Movement to the nearest neighbors.

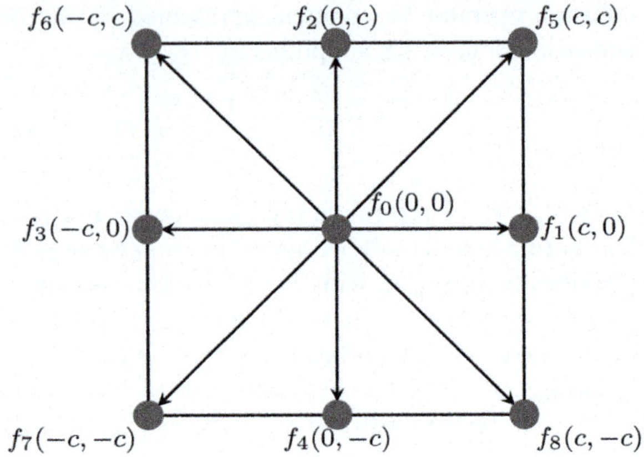

Fig. 1 Velocity vectors in the D2Q9 model

- $\boldsymbol{e}_5 = (1, 1)$, $\boldsymbol{e}_6 = (-1, 1)$, $\boldsymbol{e}_7 = (-1, -1)$, $\boldsymbol{e}_8 = (1, -1)$: Movement to the next-nearest neighbors.

Weights and distribution functions: The weights w_i associated with these velocity vectors through (4) play a fundamental role in the particle distribution calculations within the model [6]:

$$
\begin{aligned}
w_0 &= \tfrac{4}{9} && \text{for the rest particle,} \\
w_{1,2,3,4} &= \tfrac{1}{9} && \text{for particles moving toward nearest neighbors,} \\
w_{5,6,7,8} &= \tfrac{1}{36} && \text{for particles moving toward next-nearest neighbors.}
\end{aligned}
\tag{7}
$$

These weights help define the equilibrium distribution function f_{eq}, which integrates these weights with macroscopic variables like density and velocity. This function ensures compliance with the macroscopic equations of mass and momentum conservation, thereby providing accurate fluid dynamics simulations under various conditions.

1.3 Collision and Streaming Steps

In LBM, the simulation proceeds through two main steps: collision and streaming.

1.3.1 Collision Step

During this step, particles at each lattice node collide and redistribute their velocities according to the collision operator. This operator is typically based on the Bhatnagar-Gross-Krook (BGK) approximation, which simplifies the collision process by assuming a single relaxation parameter ω. The post-collision distribution function is given by [3]

$$f_i^*(\boldsymbol{x}, t) = f_i(\boldsymbol{x}, t) + \omega \left(f_{\text{eq},i}(\boldsymbol{x}, t) - f_i(\boldsymbol{x}, t) \right). \tag{8}$$

1.3.2 Streaming Step

In this step, the particles move to neighboring lattice nodes based on their velocities. The distribution function is updated as follows [3]

$$f_i(\boldsymbol{x} + \boldsymbol{e}_i \Delta t, t + \Delta t) = f_i^*(\boldsymbol{x}, t). \tag{9}$$

1.4 Boundary Conditions

Applying appropriate BC is crucial for accurately simulating fluid flow around obstacles. In LBM, several BC can be implemented easily [8]:

- **No-slip BC**: This condition is applied at solid boundaries (such as the surface of an obstacle) using the bounce-back rule. Particles that hit the boundary are reflected back along their incoming direction, ensuring that the velocity at the boundary is zero.
- **Inlet and outlet BC**: These conditions are used to specify the fluid flow at the boundaries of the simulation domain. Common approaches include specifying a constant velocity profile at the inlet and a zero-gradient condition at the outlet.

1.5 Simulation Setup

The simulation of fluid flow around a square/circular obstacle involves defining a discrete lattice grid, initializing the particle distribution functions, and iteratively applying the collision and streaming steps. The BC, including the no-slip condition on the obstacle and the periodic inlet and outlet conditions, are implemented using simple bounce-back rules. The simulation runs for a specified number of iterations, and the macroscopic properties, such as velocity and pressure fields, are computed and visualized.

The Python code generated for this project uses the D2Q9 model to simulate the fluid flow around a square/circular obstacle. The code initializes the lattice, sets up the BC, and iteratively updates the particle distribution functions. The results are visualized using velocity magnitude and pressure fields, providing insights into the flow behavior around the obstacle.

The algorithm for LBM simulations of 2D fluid flow is given in the Algorithm Box 1.

Algorithm 1: Lattice Boltzmann Method for 2D Fluid Flow

Input: Lattice size (Nx, Ny), time steps T, Reynolds number Re, initial distribution function $f_i(\boldsymbol{x}, t = 0)$, boundary conditions, obstacle type (square/circle), obstacle parameters (center (cx, cy) and size)

Output: Velocity field $\boldsymbol{u}(\boldsymbol{x}, T)$, density field $\rho(\boldsymbol{x}, T)$

Initialization:
Initialize the lattice grid with size (Nx, Ny) and Set the obstacle shape and position (square or circle);
Calculate relaxation parameter ω based on Re;
Set initial macroscopic variables: density $\rho(\boldsymbol{x}, 0)$ and velocity $\boldsymbol{u}(\boldsymbol{x}, 0)$;
Initialize the distribution function $f_i(\boldsymbol{x}, 0)$ according to equilibrium distribution function $f_i^{eq}(\rho, \boldsymbol{u})$;
for $t = 0$ **to** T **do**
    Collision Step:
    foreach *lattice node* $\boldsymbol{x} \in (Nx, Ny)$ **do**
        Calculate macroscopic variables $\rho(\boldsymbol{x}, t)$ and $\boldsymbol{u}(\boldsymbol{x}, t)$ from $f_i(\boldsymbol{x}, t)$;
        Compute equilibrium distribution $f_i^{eq}(\rho, \boldsymbol{u})$;
        Update distribution function using BGK approximation:

$$f_i(\boldsymbol{x}, t + \Delta t) = f_i(\boldsymbol{x}, t) - \omega \left(f_i(\boldsymbol{x}, t) - f_i^{eq}(\rho, \boldsymbol{u}) \right)$$

    end
    Streaming Step:
    foreach *lattice direction i* **do**
        Move the distribution function to neighboring nodes:

$$f_i(\boldsymbol{x} + \boldsymbol{c}_i \Delta t, t + \Delta t) = f_i(\boldsymbol{x}, t + \Delta t)$$

    end
    Apply Boundary Conditions:
    Apply appropriate boundary conditions (e.g., bounce-back for walls, periodic, outflow conditions etc.) on $f_i(\boldsymbol{x}, t + \Delta t)$;
    Update Macroscopic Quantities:
    Calculate macroscopic quantities $\rho(\boldsymbol{x}, t + \Delta t)$ and $\boldsymbol{u}(\boldsymbol{x}, t + \Delta t)$;
end
Output:
Return the final velocity field $\boldsymbol{u}(\boldsymbol{x}, T)$ and pressure field $\rho(\boldsymbol{x}, T)$;

1.6 Overview of the Python Code

Ideally, the program generated by AI for this project should use the D2Q9 model to simulate fluid flow around a square or circular obstacle and consist of the following key components: [5]

1. **Initialization**: The lattice grid should be defined, and the particle distribution functions should be initialized.
2. **Collision and streaming**: The code should iteratively apply the collision and streaming steps to update the particle distribution functions.
3. **Boundary conditions**: No-slip BC should be applied using the bounce-back rule. Inlet and outlet conditions should be specified to maintain the flow.
4. **Visualization**: The macroscopic properties, such as velocity magnitude and pressure fields, should be computed and visualized to analyze the flow behavior.

The AI-generated code demonstrates the potential of artificial intelligence in assisting with complex computational tasks. By providing a detailed prompt, the AI was able to generate a functional code for simulating fluid flow using the LBM.

This project uses AI, specifically GPT-4o, to assist in coding the LBM for simulating fluid flow around an obstacle. The goal is to assess the capability of AI in generating Python code for this complex task. The subsequent sections will present the specific problem to be solved, the AI-generated prompt, the resulting code, outcomes, its verification and some remarks about the GPT-4o.

By looking at the intersection of AI and CFD, this report may aim to highlight the strengths and limits of AI-assisted coding in scientific research.

2 Prompt

2.1 Problem Statement Definition

The simulation involves modeling fluid flow around an obstacle within a 2D rectangular domain using the LBM. The domain is defined with the following parameters:

- **Domain size**: The computational domain has a width nx and a height ny. Here 1 lattice spacing equals 1 mm i.e., there will be nx lattice nodes in x direction for nx mm width and ny lattice nodes in y direction for ny mm height.
- **Coordinate system**: The origin (0,0) is located at the bottom-left corner of the domain. The x-axis extends horizontally to the right, and the y-axis extends vertically upwards.
- **Obstacle**: The obstacle can be either a square or a circle, positioned within the domain based on user input.
  - For a **square obstacle**, the user specifies the x and y coordinates of the center (cx, cy) and the side length.

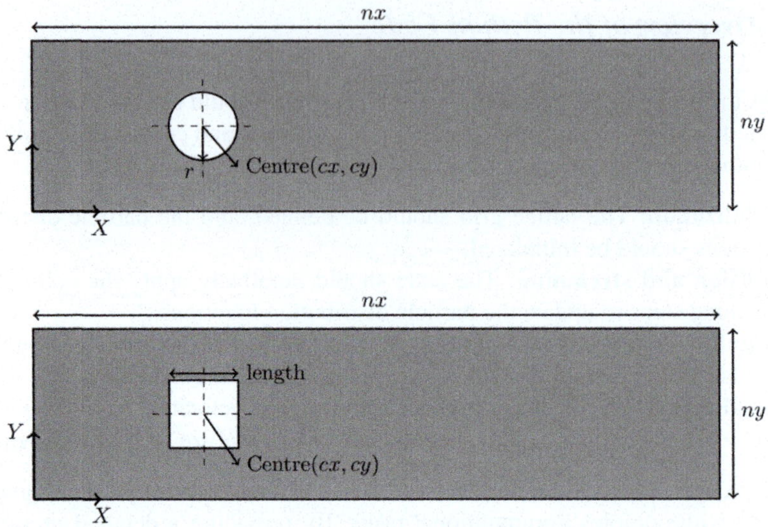

Fig. 2 Computational domain with (top) a circular and (bottom) a square obstacle

- For a **circular obstacle**, the user specifies the x and y coordinates of the center (cx, cy) and the radius.

- **Discretization**: The lattice nodes are evenly spaced, forming a grid over the domain. The flow simulation uses these discrete points to model fluid dynamics.

Figure 2 sketches the simulation domain with exemplary obstacles.

For definiteness in the coding assignment, the parameters are chosen as mentioned in Table 1.

Table 1 Parameters chosen for the fluid simulation study with obstacles

Parameters	Values/Types	Units
Domain width (nx)	400	mm
Domain height (ny)	100	mm
Obstacle type	Circular/Square	
Obstacle center (cx, cy)	(100, 50)	(mm, mm)
Radius of circle	25	mm
Length of square	40	mm

(continued)

The final, working prompt is given in Fig. 3. It is used for generating the Code Listing 1.

Table 1 (continued)

Parameters	Values/Types	Units
LBM cell size	1	mm
time step size Δt	1	s
Total time steps	4000	
Reynolds number (Re)	50	
Upstream velocity $\boldsymbol{u}$	0.04	mm/s
Model type	Bounce-back	
Obstacle BC	No-slip	
Output	Velocity, Pressure and streamline plots	

Act as a computational material scientist and Python expert.
Write a comprehensive Python script using the Lattice Boltzmann Method to simulate 2D flow around an obstacle. The script should accomplish the following tasks:

- Initialization:
  - Define a 2D rectangular computational domain with a width nx of 400 lattice nodes and a height ny of 100 lattice nodes.
  - Initialize the lattice grid and the particle distribution functions for each lattice node.
  - Ask the user to choose the type of obstacle (square or circular) and specify the dimensions and position:
    - For a square obstacle, prompt for the x and y coordinates of the center (cx, cy) and the side length.
    - For a circular obstacle, prompt for the x and y coordinates of the center (cx, cy) and the radius.
    - Ensure the user inputs are validated to prevent the obstacle from being positioned outside the domain.
  - Set Reynolds number $Re = 50$ to model moderate flow conditions. Use the lattice velocity $u_{\mathrm{LB}} = 0.04$ as a characteristic velocity.

Fig. 3 Prompt for simulating fluid flow around an obstacle

- Functions and Simulation Loop:
  - Implement the LBM using the D2Q9 model, including the following steps:
    - Collision Step: Use the BGK approximation for the collision operator:

      $$f_{\text{out}} = f_{\text{in}} - \omega \cdot (f_{\text{in}} - f_{\text{eq}})$$

      where ω is the relaxation parameter, and f_{eq} is the equilibrium distribution function.
    - Streaming Step: Move the particles to neighboring lattice nodes based on their velocities.
    - Boundary Conditions: Apply no-slip BC at the obstacle using the bounce-back rule. Specify inlet and outlet conditions to maintain the flow.
  - Run the simulation for iterations of 4000 time steps, updating the particle distribution functions iteratively.
- Macroscopic Variables:
  - Calculate macroscopic variables such as density ρ, velocity ($\boldsymbol{u}$), and pressure at each lattice node.
- Expected Output:
  - Use Matplotlib to visualize the results:
    - Plot the velocity magnitude and pressure field.
    - Create a denser streamline plot to depict the flow direction profile.
    - Plot the velocity profile at each time step to create an animation of the flow.

Ensure the script is well-documented and modular, with clear function definitions and comments explaining each step of the process.

Fig. 3 (continued)

3 Code Listing

The final code version generated by GPT-4o is given in the Code Listing 1.

```python
import numpy as np
import matplotlib.pyplot as plt
from matplotlib import cm
import pandas as pd

# Function to get obstacle parameters from the user
def get_obstacle_parameters():
    obstacle_type = input("Enter the type of obstacle (square/circle): ").strip().lower()
    if obstacle_type == "square":
        return get_square_parameters()
    elif obstacle_type == "circle":
        return get_circle_parameters()
    else:
        raise ValueError("Invalid obstacle type. Please enter 'square' or 'circle'.")

# Function to get parameters for a square obstacle from the user
def get_square_parameters():
    while True:
        try:
            cx = int(input("Enter the x-coordinate of the square center (cx): "))
            cy = int(input("Enter the y-coordinate of the square center (cy): "))
            side = int(input("Enter the side length of the square: "))
            # Ensure the square is within the simulation domain
            if cx - side // 2 < 0 or cx + side // 2 >= nx or cy - side // 2 < 0 or cy + side // 2 >= ny:
                raise ValueError("Square parameters are outside the region or not feasible.")
            return "square", cx, cy, side
        except ValueError as e:
            print(e)

# Function to get parameters for a circular obstacle from the user
def get_circle_parameters():
    while True:
        try:
            cx = int(input("Enter the x-coordinate of the circle center (cx): "))
            cy = int(input("Enter the y-coordinate of the circle center (cy): "))
            r = int(input("Enter the radius of the circle (r): "))
            # Ensure the circle is within the simulation domain
```

```python
            if cx - r < 0 or cx + r >= nx or cy - r < 0 or cy + r >= ny:
                raise ValueError("Circle parameters are outside the region or not feasible.")
            return "circle", cx, cy, r
        except ValueError as e:
            print(e)

###### Flow definition ##########################################################
maxIter = 4000  # Total number of time iterations.
nx, ny = 400, 100  # Number of lattice nodes in the x and y directions.
ly = ny - 1  # Height of the domain in lattice units.
Re = 50.0  # Reynolds number.
uLB = 0.04  # Velocity in lattice units.

# Get obstacle parameters from the user
obstacle_type, cx, cy, size = get_obstacle_parameters()
# Calculate viscosity in lattice units based on Reynolds number and characteristic length (size)
nulb = uLB * size / Re
# Calculate relaxation parameter for the Lattice Boltzmann method
omega = 1. / (3. * nulb + 0.5)

###### Lattice Constants #####################################################
# Define the discrete velocity set for D2Q9 model (9 velocities)
v = np.array([[1, 1], [1, 0], [1, -1], [0, 1], [0, 0], [0, -1], [-1, 1], [-1, 0], [-1, -1]])
# Define the weights for each velocity direction
t = np.array([1./36., 1./9., 1./36., 1./9., 4./9., 1./9., 1./36., 1./9., 1./36.])

# Define columns for streaming step (used for shifting the distribution functions)
col1 = np.array([0, 1, 2])
col2 = np.array([3, 4, 5])
col3 = np.array([6, 7, 8])

###### Function Definitions ##################################################
# Function to compute macroscopic variables (density and velocity) from distribution functions
def macroscopic(fin):
    """
    Compute macroscopic variables (density rho and velocity u) from the distribution functions (fin).
    Args:
    fin (ndarray): Distribution functions.
    Returns:    rho (ndarray): Density field.
                u (ndarray): Velocity field."""
    # Compute density as the sum of distribution functions
```

```
    rho = np.sum(fin, axis=0)
    # Initialize velocity array with zeros
    u = np.zeros((2, nx, ny))
    # Compute velocity as the weighted sum of distribution
    ↪ functions
    for i in range(9):
        u[0, :, :] += v[i, 0] * fin[i, :, :]
        u[1, :, :] += v[i, 1] * fin[i, :, :]
    # Normalize velocity by density
    u = u / rho
    return rho, u

# Function to compute the equilibrium distribution function
def equilibrium(rho, u):
    """
    Compute the equilibrium distribution function based on
    ↪ density and velocity.
    Args:
    rho (ndarray): Density field.
    u (ndarray): Velocity field.
    Returns:  feq (ndarray): Equilibrium distribution functions.
    ↪    """
    # Compute the square of the velocity magnitude
    usqr = 3. / 2. * (u[0]**2 + u[1]**2)
    # Initialize equilibrium distribution function array
    feq = np.zeros((9, nx, ny))
    # Compute equilibrium distribution function for each velocity
    ↪  direction
    for i in range(9):
        cu = 3. * (v[i, 0] * u[0, :, :] + v[i, 1] * u[1, :, :])
        feq[i, :, :] = rho * t[i] * (1. + cu + 0.5 * cu**2 - usqr
    ↪ )
    return feq

###### Setup: obstacle and velocity inlet with perturbation
    ↪ ########
# Function to create a mask for a square obstacle
def obstacle_fun_square(x, y):
    return np.logical_and(np.abs(x - cx) <= size // 2, np.abs(y -
    ↪  cy) <= size // 2)

# Function to create a mask for a circular obstacle
def obstacle_fun_circle(x, y):
    return (x - cx)**2 + (y - cy)**2 <= size**2

# Create the obstacle mask based on user input
if obstacle_type == "square":
    obstacle = np.fromfunction(obstacle_fun_square, (nx, ny))
else:
    obstacle = np.fromfunction(obstacle_fun_circle, (nx, ny))

# Function to initialize the velocity field with a slight
    ↪ perturbation
```

```
def inivel(d, x, y):
    """
    Initialize the velocity field with a slight perturbation to trigger instabilities.
    Args:
    d (int): Dimension index (0 for x-direction, 1 for y-direction).
    x (ndarray): x-coordinates of the grid points.
    y (ndarray): y-coordinates of the grid points.
    Returns:
    velocity (ndarray): Initial velocity field.
    """
    return (1. - d) * uLB * (1. + 1.e-4 * np.sin(y / ly * 2. * np.pi))

# Initialize the velocity field with the perturbation
vel = np.fromfunction(inivel, (2, nx, ny))

# Initialize the distribution functions at equilibrium with the initial velocity
fin = equilibrium(1., vel)

###### Main time loop ##########################################################
for time in range(maxIter):
    # Right wall: outflow condition (copy the values from the second last column to the last column)
    fin[col3, -1, :] = fin[col3, -2, :]

    # Compute macroscopic variables (density and velocity)
    rho, u = macroscopic(fin)

    # Left wall: inflow condition (impose velocity and compute density)
    u[:, 0, :] = vel[:, 0, :]
    rho[0, :] = 1. / (1. - u[0, 0, :]) * (np.sum(fin[col2, 0, :], axis=0) + 2. * np.sum(fin[col3, 0, :], axis=0))

    # Compute equilibrium distribution function based on updated macroscopic variables
    feq = equilibrium(rho, u)
    fin[[0, 1, 2], 0, :] = feq[[0, 1, 2], 0, :] + fin[[8, 7, 6], 0, :] - feq[[8, 7, 6], 0, :]

    # Collision step: relaxation towards equilibrium
    fout = fin - omega * (fin - feq)

    # Bounce-back condition for obstacle: reflect distribution functions
    for i in range(9):
        fout[i, obstacle] = fin[8 - i, obstacle]

```

```
    # Streaming step: propagate the distribution functions to
    ↪ neighboring nodes
    for i in range(9):
        fin[i, :, :] = np.roll(np.roll(fout[i, :, :], v[i, 0],
    ↪ axis=0), v[i, 1], axis=1)

    # Visualization every 100 iterations
    if time % 100 == 0:
        plt.clf()
        # Plot velocity magnitude
        plt.imshow(np.sqrt(u[0]**2 + u[1]**2).transpose(), cmap=
    ↪ cm.Reds)
        plt.colorbar()
        plt.title(f'Velocity magnitude at time step {time}')
        plt.pause(0.1)

###### Post-Processing
    ↪ ############################################################
# Compute final macroscopic variables (density and velocity) from
    ↪ distribution functions
rho, u = macroscopic(fin)

# Set velocity to zero inside the obstacle (no flow inside the
    ↪ obstacle)
u[0, obstacle] = 0
u[1, obstacle] = 0

# Pressure field calculation (assuming ideal gas law: p = rho *
    ↪ RT, with RT = 1)
pressure = rho
pressure[obstacle] = np.nan  # Mask the pressure inside the
    ↪ obstacle

###### Plotting
    ↪ ##############################################################
# Compute velocity magnitude field
velocity_magnitude = np.sqrt(u[0]**2 + u[1]**2)
velocity_magnitude[obstacle] = np.nan # Mask the velocity
    ↪ magnitude inside the obstacle

plt.figure(figsize=(12, 6))
plt.subplot(121)
# Plot velocity magnitude
plt.imshow(velocity_magnitude.transpose(), cmap=cm.viridis)
cbar=plt.colorbar(location='bottom')
cbar.ax.tick_params(labelsize=13)
plt.title('Velocity Magnitude(m/s)',size='16')
plt.xticks(fontsize=13)
plt.yticks(fontsize=13)

plt.subplot(122)
# Plot pressure field
plt.imshow(pressure.transpose(), cmap=cm.viridis)
```

```
cbar=plt.colorbar(location='bottom')
cbar.ax.tick_params(labelsize=13)
plt.title('Pressure(Pa)',size='16')
plt.xticks(fontsize=13)
plt.yticks(fontsize=13)

plt.tight_layout()
plt.show()

###### Streamline Plot
    ↪ ##############################################################
# Plot streamlines to visualize the flow direction
plt.figure(figsize=(12, 6))
plt.streamplot(np.arange(nx), np.arange(ny), u[0].transpose(), u
    ↪ [1].transpose(), color=velocity_magnitude.transpose(),
    ↪ density=2, cmap=cm.viridis)
cbar=plt.colorbar(location='bottom')
cbar.ax.tick_params(labelsize=18)
plt.title('Streamlines plot',size='23')
plt.xticks(fontsize=18)
plt.yticks(fontsize=18)
plt.gca().set_aspect('equal')  # Set aspect ratio to be equal
plt.tight_layout()
plt.show()
```

Listing 1 Output by GPT-4o for solving fluid flow around an obstacle in 2D

The output of the GPT-4o does not guarantee that the code generated will work without any flaws. More discussion on the understanding of the GPT-4o is mentioned in Sect. 5.

4 Test for Verification

Verification is crucial to ensure the accuracy and reliability of computational simulations. It involves comparing the results of a simulation with analytical solutions or experimental data to confirm that the model behaves as expected.

To verify the results obtained from the Python code, a similar simulation is performed using ANSYS Fluent. The ANSYS simulation setup involves defining the same computational domain and obstacle dimensions, applying appropriate BC, and running the simulation as in the Python LBM code.

The fluid flow is modeled using the Navier-Stokes equations, solved with the Finite Volume Method (FVM). The domain is discretized using a structured grid, where the grid cells are organized in a regular pattern. A higher mesh density is used to capture accurate boundary layer effects around the obstacle while keeping the grid resolution in the interior of the channel coarser thus balancing efficiency with accuracy.

The element size implemented in the simulation is $h = 2.0616 \times 10^{-5}$mm. Similar to the LBM counterpart in this work, a constant time-stepping method is employed with a fixed time step size of $\Delta t = 0.04$ s.

The BCs include a uniform inflow with a slight perturbation an outflow condition at the domain's exit, and no-slip conditions on the obstacle surface to simulate the interaction between the fluid and the solid boundary.

The results from ANSYS Fluent provide a benchmark to compare and validate the results obtained from the AI-generated Python code. The simulation results, including velocity and pressure fields, are compared against the Python-based Lattice Boltzmann Method (LBM) simulation in Sect. 4.1.

4.1 Comparison

To verify the results, we compare the velocity, the streamline, and the pressure profiles (Figs. 4 and 5) as well as the midline of the geometry ($ny = 50$ mm) for both the Python code and ANSYS simulations (Fig. 6). Velocity data is extracted from both the Python code and the ANSYS simulations for this purpose. The comparison in Fig. 6a and b for the x-velocity and y-velocity profiles indicate excellent agreement of the present code with the results using the commercial solver.

The profiles obtained from the Python code and ANSYS Fluent simulations show a high degree of agreement. The streamline patterns and velocity magnitude plots exhibit similar flow characteristics around the obstacle. Specifically:

- **Streamline plots**: The results of both simulation frameworks (Fig. 4a and c) exhibit the expected vortex shedding behind the obstacle, indicating that the flow separation and recirculation regions are captured accurately.
- **Velocity contour plots**: They agree between the present simulation results with the commercial solver with respect to the velocity distributions and the maxima (Fig. 4a and c).
- **Velocity component profiles**: The x-velocity and y-velocity profiles at the midline of the geometry (Fig. 6) are in excellent agreement for the present Python code and the commercial solver, which further strongly suggests the correctness of the GPT4o-generated code.

Despite minor discrepancies, the overall agreement between the two sets of results demonstrates that the GPT-4o-generated code accurately simulates the fluid flow around the obstacle. The close match between the streamline patterns, velocity magnitude distributions, and velocity profiles indicates that the Python code of this work performs on par with the sophisticated commercial software ANSYS Fluent.

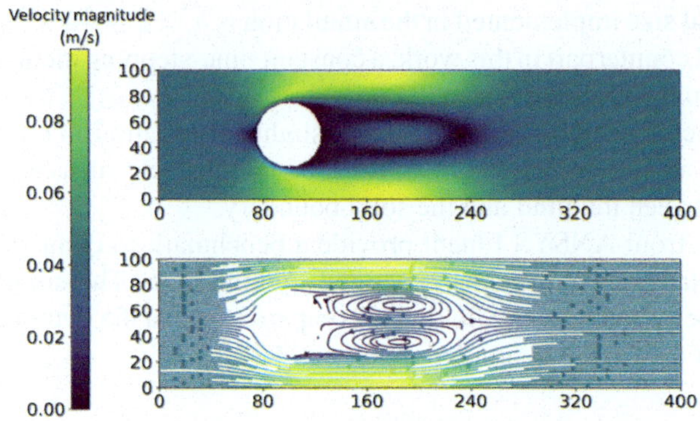

(a) Velocity (top) and Streamline (bottom) plot for the circular obstacle

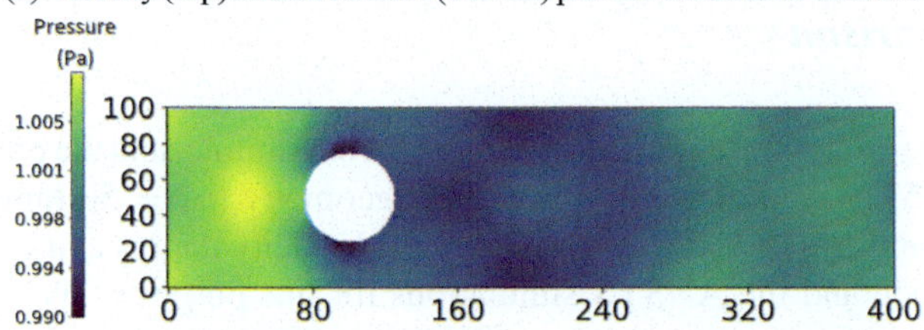

(b) Pressure distribution plot for the circular obstacle

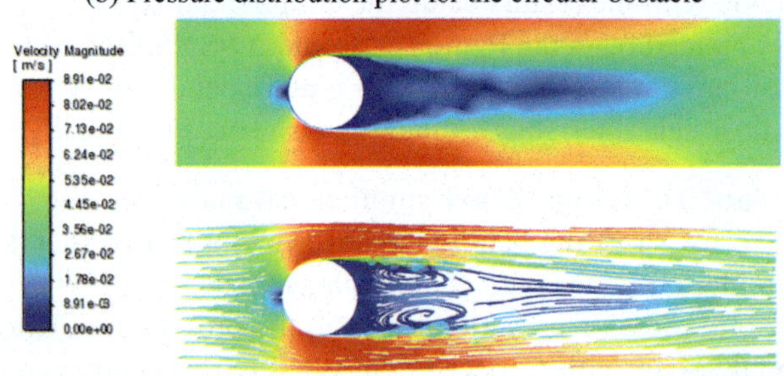

(c) Velocity (top) and streamline (bottom) plots from the ANSYS Fluent simulation

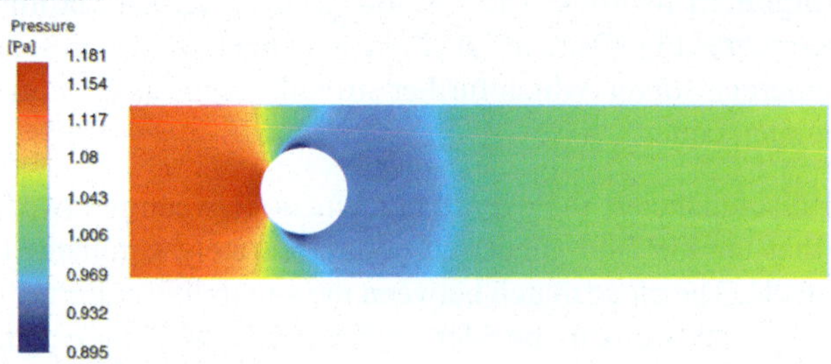

(d) Pressure distribution plot for the circular obstacle from the ANSYS Fluent simulation

Fig. 4 Simulation results from python code **a**, **b** with a comparison to a reference solution **c**, **d** from a commercial solver for the circular obstacle

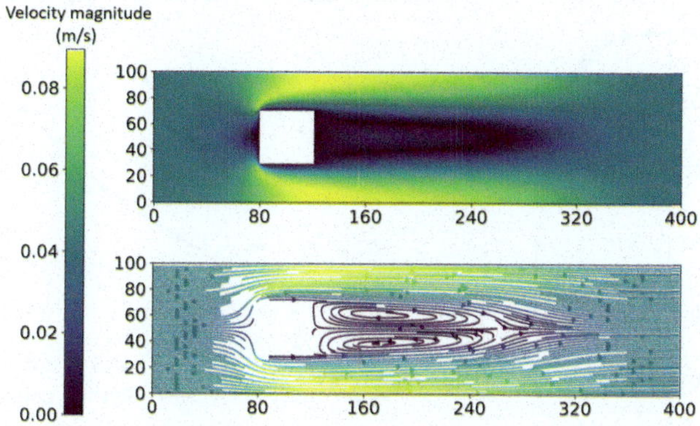

(a) Velocity (top) and Streamline (bottom) plot for the square obstacle

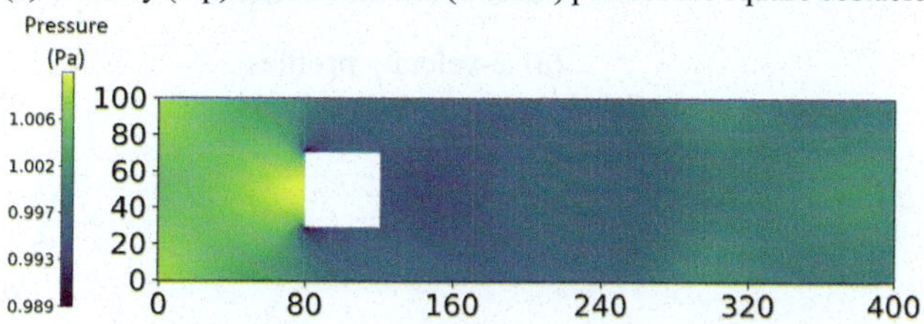

(b) Pressure distribution plot for the square obstacle

Fig. 5 Simulation results from python code **a**, **b** for the square obstacle

4.2 Discussion of Flow Phenomena at Reynolds Number 50 Versus 300

Beyond the comparison carried out above, the characteristics of the flow around the circular obstacle at a Reynolds number 50 are briefly analyzed and set into comparison with additional results for Re = 300, which reveals interesting fluid dynamics phenomena. These can be categorized into three distinct regions:

4.2.1 Flow in Front of the Obstacle (Upstream)

- **Flow Deceleration and Stagnation Point**: As the fluid approaches the circular obstacle, it begins to decelerate due to the obstruction. Directly in front of the obstacle, a stagnation point forms where the flow velocity reduces to zero. At this point, the fluid is diverted around the obstacle.
- **Pressure Increase**: The deceleration of fluid near the stagnation point causes a rise in pressure in front of the obstacle, following Bernoulli's principle. The flow

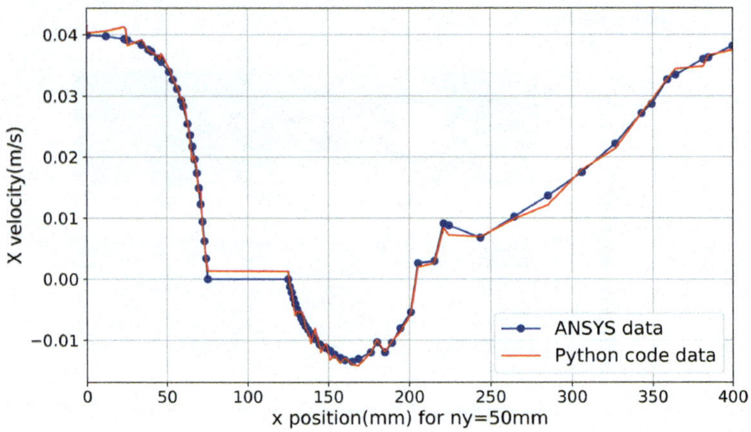

(a) x-velocity profiles

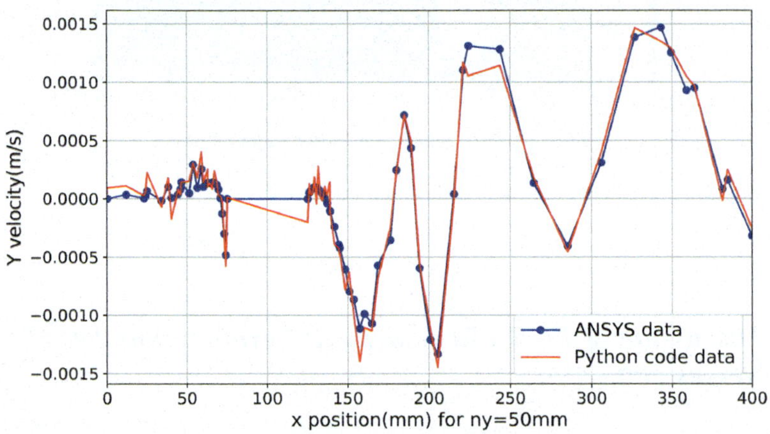

(b) y-velocity profiles

Fig. 6 Comparison of velocity profiles at the mid-line ($ny = 50$ mm) for the circular obstacle between the present python simulation result and a reference solution from the commercial solver ANSYS

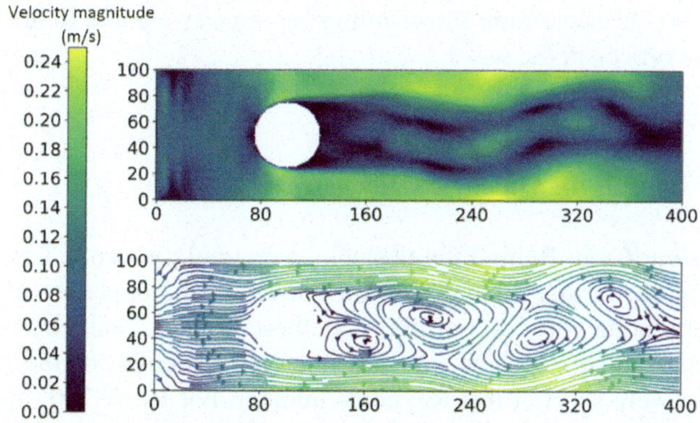

(a) Velocity contour plot (top) and streamline plot (bottom)

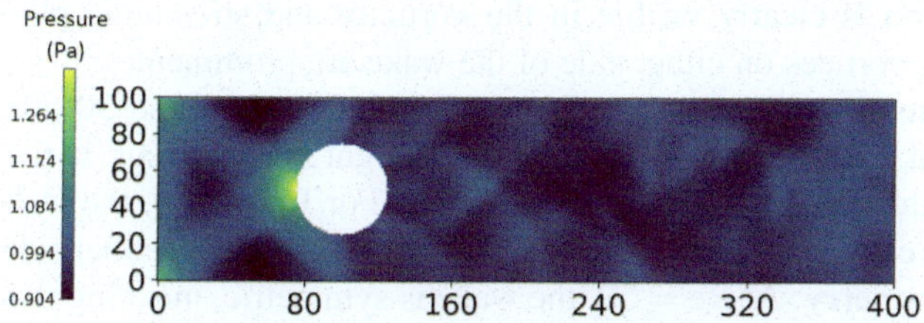

(b) Pressure distribution contour plot

Fig. 7 Simulation results for the circular obstacle at Re = 300

lines are closer together, indicating a pressure build-up. Note that for Re = 50 the pressure maximum is in some distance to the obstacle, Fig. 4b, for Re = 300 right at the obstacle, Fig. 7b.

4.2.2 Flow Passing the Obstacle (Along the Sides)

- **Boundary Layer Development**: As the fluid moves around the obstacle, a boundary layer forms along the surface of the obstacle. This boundary layer is initially laminar due to the low Reynolds number but starts thickening as it progresses along the sides.
- **Flow Separation**: Given that the Reynolds number is around 50, flow separation occurs on the sides of the obstacle. The flow cannot remain attached to the surface of the obstacle as it curves around the sides. This results in the boundary layer separating from the surface at some point, usually on the rear half of the obstacle.
- **Symmetry in Separation**: At Re = 50, the separation is symmetric in that the separation points on both sides of the obstacle occur at roughly the same position.

For Re $=300$, the separation shows minor asymmetry leading to the formation of alternating vortices in the wake.

4.2.3 Flow Behind the Obstacle (Downstream)

- **Recirculation Zone**: Behind the obstacle, a recirculation zone forms due to the flow separation. In this region, the fluid moves in the opposite direction to the main flow, creating vortices. For Re $=50$, these vortices –known as *recirculating eddies* or *vortex pairs*– show high symmetry. The length of the re-circulation zone is relatively short at this Reynolds number. For Re $=300$, in contrast, the recirculation zone exhibits alternating vortices. The flow enters a periodic vortex shedding regime, leading to the formation of a *von Kármán vortex street*. This phenomenon is clearly visible in the vorticity and streamline plots, where the alternating vortices on either side of the wake are prominent.
- **Vortex Shedding Onset**: At a Reynolds number of around 50, the flow is in a transitional regime where vortex shedding might start to occur; in the present case however, there is no clear vortex shedding. For Re $=300$, the wake behind the obstacle is characterized by alternating vortices forming on either side of the wake.
- **Wake Symmetry**: At Re $=50$, the wake is symmetric, meaning that the vortices are of similar size and strength on both sides of the obstacle. More downstream in some distance to the obstacle the flow turns back to laminar. For Re $=300$ the wake is inherently asymmetric due to the unsteady vortex shedding. The vortices alternate in strength and size, which is characteristic of the vortex shedding process in this Reynolds number range. **Transition to Turbulence**: The flow exhibits characteristics of transition towards turbulence in the wake region. The alternating vortices begin to interact with each other, and their dynamics contribute to the development of complex, unsteady flow patterns further downstream

These observations highlight the significant impact of increasing the Reynolds number on the flow behavior around the obstacle. The flow becomes more unsteady, with prominent alternating vortex shedding and complex wake dynamics, indicating the growing dominance of inertial forces over viscous forces compared to Re $=50$.

5 Discussion

This section discusses various aspects of the project, including errors and omissions, completeness, reproducibility, and the lessons learned during the process.

5.1 Errors and Omissions

During the code generation process, various errors and omissions are encountered, which can be categorized into major and minor issues.

- **Major Issues**:
  - Index mismatching errors: These errors were frequent and challenging to recognize, often leading to the code failing to execute properly.
  - Non-reproducibility of code: The generated code varies significantly with each attempt, even with the same prompt. This inconsistency makes it difficult to achieve reliable results.
  - Incomplete outputs: On several occasions, the code runs without errors but fails to produce any output, resulting in blank profiles.
- **Minor Issues**:
  - Missing details: Despite providing detailed prompts, GPT-4o sometimes misses implementing simple yet crucial details.

5.2 Completeness and Short Term Memory Capacity

GPT-4o demonstrates a good capacity to implement details as specified in the prompt. However, it is noted that while detailed prompts are necessary, extremely detailed prompts can cause GPT-4o to forget some of the details, leading to incomplete implementation. It is crucial to strike a balance in the level of detail provided in the prompts to ensure the best results.

5.3 Reproducibility as a Signature of Reliability

Unfortunately, the reproducibility of the code using the same prompt is not assured. Several observations are noted during the generation of the code and the formulation of the prompt:

- GPT-4o generates different codes for each trial, with many instances resulting in compilation errors such as array dimension mismatches.
- Allocating a persona to GPT-4o which is an expert in the respective domain is beneficial to a large extent. Additionally, when dealing with specialized subjects, providing GPT-4o with useful principles and equations enhances the accuracy of the results.

- The previous history of the session fed to GPT-4o plays a critical role in generating further results more accurately according to the need. Therefore, the same assignment, even with the same formulation, may not produce identical outputs due to the influence of the session history.

5.4 Learned Lessons

Several valuable lessons were learned during the project:

- The importance of providing clear, detailed, and structured prompts to GPT-4o to improve the quality and accuracy of the generated code.
- Understanding the limitations of GPT-4o in terms of reproducibility and the influence of session history on the generated outputs.
- Recognizing the necessity of human intervention to refine and validate the results, ensuring the accuracy and reliability of the simulations.
- Appreciating the potential of AI-assisted coding in generating complex algorithms, while also acknowledging the need for human expertise to refine and verify the results.

Overall, this project highlights both the strengths and limitations of using AI tools like GPT-4o for CFD simulations. The experience underscores the importance of combining AI capabilities with human expertise to achieve reliable and accurate results.

6 Conclusion

This chapter demonstrated the capability of GPT-4o to generate functional Python code for simulating fluid flow around an obstacle using the LBM. The generated code was verified in a benchmark simulation against the results of the commercial solver ANSYS Fluent; the comparison indicates a high degree of agreement and confirms the accuracy of the AI-assisted coding approach.

While GPT-4o proved to be a powerful tool in generating complex CFD algorithms, human intervention was crucial for refining, debugging, and verifying the results in the present work. In particular we pointed at challenges related to reproducibility and error handling, emphasizing the need for clear, balanced prompts and human oversight. Despite these challenges, the verification against ANSYS Fluent demonstrated that GPT-4o-generated code can perform on par with advanced commercial software, underscoring the viability of GPT-4o-assisted coding for scientific research and engineering applications.

Overall, while GPT-4o cannot completely replace human expertise, it can significantly speed up the coding process, making it a valuable tool for enhancing efficiency and innovation in computational simulations.

References

1. T. Krüger, H. Kusumaatmaja, A. Kuzmin, O. Shardt, G. Silva, and E. M. Viggen, "The Lattice Boltzmann Method: Principles and Practice," Springer, 2017. https://doi.org/10.1007/978-3-319-44649-3.
2. S. Chen and G. D. Doolen, "Lattice Boltzmann Method for Fluid Flows," *Annual Review of Fluid Mechanics*, vol. 30, pp. 329–364, 1998. https://doi.org/10.1146/annurev.fluid.30.1.329.
3. J. Yojina, W. Ngamsaad, N. Nuttavut, D. Triampo, Y. Lenbury, P. Kanthang, S. Sriyab, and W. Triampo, "Investigating flow patterns in a channel with complex obstacles using the lattice Boltzmann method," *Journal of Mechanical Science and Technology*, vol. 24, no. 10, pp. 2025–2034, 2010.
4. W. Abassi, F. Aloui, S. Ben Nasrallah, and J. Legrand, "Lattice Boltzmann Method Used to Simulate an Unsteady Flow Around an Obstacle in Laminar Regime," in *ASME/JSME/KSME 2011 Joint Fluids Engineering Conference*, Hamamatsu, Japan, July 2011.
5. P. Mocz, "latticeboltzmann-python: Lattice Boltzmann simulation in Python," GitHub repository, 2023. [Online]. Available: https://github.com/pmocz/latticeboltzmann-python. [Online accessed: June-2024].
6. A. Majumder, V. Kariwala, S. Ansumali, and A. Rajendran, "Lattice Boltzmann method for population balance equations with simultaneous growth, nucleation, aggregation and breakage," *Chemical Engineering Science*, vol. 69, no. 1, pp. 316–328, 2012. https://doi.org/10.1016/j.ces.2011.10.051.
7. D. Heubes, A. Bartel, and M. Ehrhardt, *An Introduction to the Lattice Boltzmann Method for Coupled Problems*, Preprint BUW-IMACM 11/28, Institute of Mathematical Modelling, Analysis and Computational Mathematics (IMACM), Bergische Universität Wuppertal, December 2011.
8. Z. Guo and C. Shu, *Lattice Boltzmann Method and its Applications in Engineering*. World Scientific Publishing Co. Pte. Ltd., 2013.

Conclusions

Learned Lessons—Recommendations

Bernhard Eidel, Rahul Narkhede, and Aagashram Neelakandan

Abstract This chapter condenses the lessons learned in different chapters of this book into recommendations for prompt engineering of Python programming tasks for the chatbot GPT-4/4o in the areas of computational materials and mechanics. Beyond, detailed recommendations for code verification are provided.

1 Responsible Usage of GPT-4 in Coding Tasks

Expert knowledge in the areas in which the chatbot is used as a programming assistant is absolutely necessary for responsible use. This refers to the domain of modeling problems in nature and the sciences by differential equations and to numerical methods employed for their solution. It no less refers to the field of professional programming.

Hence, this expert knowledge must be readily available before the prompt is shaped—and will certainly be of great help to craft the prompt. This expert knowledge equally is required in rigorous tests of the code implying verification tests as well as validation tests. In this book we had restricted to verification, which is, briefly, the assessment of the accuracy of the code and its solution to a computational model by comparison with known solutions. Validation is the assessment of the accuracy of a computational simulation by comparison with experimental data.

While GPT-4 proved to be a powerful tool in generating complex algorithms, human intervention was throughout necessary in all chapters of this book.

B. Eidel (✉) · R. Narkhede · A. Neelakandan
M5, IMFD, TUBAF, Freiberg, Germany
e-mail: bernhard.eidel@imfd.tu-freiberg.de

R. Narkhede
e-mail: rahul-vishnu.narkhede@student.tu-freiberg.de

A. Neelakandan
e-mail: aagashram.neelakandan@student.tu-freiberg.de

© The Author(s), under exclusive license to Springer Nature Switzerland AG 2025
B. Eidel (ed.), *GPT for Python-Coding in Computational Materials Science and Mechanics*, Studies in Computational Intelligence 1198,
https://doi.org/10.1007/978-3-031-85470-5_11

2 Tangible Advice for Prompt Engineering

In working with GPT-4 to generate Python code for computational problems, particularly in computational materials science, solid mechanics, wave propagation, and fluid mechanics, we have distilled several key lessons and strategies that can enhance the effectiveness of AI interactions. The following are detailed recommendations for prompt engineering that have proven valuable in eliciting high-quality responses from GPT-4. We include links to the general characteristics of LLMs and their performance in Chatbots.

In a book context, where the interaction with GPT-4 is limited to one or a few well-constructed prompts, it becomes essential to design each prompt carefully to ensure it leads to high-quality code generation. Multiple iterative prompts or feedback loops are often impractical, so the focus should be on refining each prompt to be as effective as possible from the outset.

1. **Persona Assignment: Generating Context**
   One effective technique is the strategic use of *persona assignment*, where GPT-4 is attributed a specific role or context. By assigning a persona as an expert with knowledge in, e.g., computational materials science with particular expertise in grain growth, the generated solutions are more aligned with user expectations. This helps GPT-4 frame its answers appropriately, adding a layer of sophistication. This approach provides context, ensuring that the output is relevant both in terms of solution and explanation. The *persona assignment* can be also used to tailor the responses generated by the chatbot to follow a certain format and take certain general instructions into consideration.
2. **Be Explicit in What You Need: Specify Language, Output Format, or Techniques**
   Clarity in prompts leads to clarity in responses. It is essential to specify the coding language, the methods or libraries to be used, and the desired output format. Some measures to ensure clarity in the prompts are listed here:

   - Provide equations or relations in a LaTeXformat to ensure accurate interpretation by GPT.
   - Mention the datatype of the inputs and expected outputs, with examples. If the inputs are images or any other form of data, mention the file format, e.g. '.jpg' or '.tif'.
   - Clearly mention the expected dimensions of matrices and vectors.
   - If a term has the possibility of being interpreted differently by GPT, then briefly specify the desired meaning in the prompt.

   This ensures the solution is tailored precisely to the problem's requirements, avoiding irrelevant approaches.
3. **Iterative Refinement Process**
   From our experience in almost all topical chapters, the first iteration of a prompt was very rarely the final one, yielding the best code output. In cases of complex

problems requiring decomposition of the problem into multiple steps, it is suggested to start a new iteration in a new chat.
One can also use GPT-4 to create a prompt using the solution from the previous iteration or briefly explain the desired output. This may act as a good starting point for subsequent prompt refinement.
By refining the prompt before the final submission, the output will be more complete, reducing the need for further iterations.

4. **Problem Decomposition**
   When addressing complex tasks, it is beneficial to break the problem into smaller, manageable components. Instead of requesting a complete solution at once, decomposing the task improves both clarity and output quality. Ideally, problems should be broken down into reasonable and logical parts, e.g., in machine learning, the problem can be split into the steps of the workflow, i.e., model inputs, model creation, training, and testing. GPT-4 can also be prompted to tackle a complex problem in a step-wise manner by adding the line "Let's think step-by-step" to invoke the *chain of thought* technique of prompt engineering. If the prompt-response chat becomes too long, it is recommended to provide direct references to previous responses or statements by simply adding them to the current prompt. Thus, the issues rendered by a limited memory can be overcome.
   This step-by-step approach structures the interaction, guiding GPT-4 to handle complex problems methodically.
5. **Embedded Explanation and Commentary: Chain of Thought in a Single Prompt** To enhance the clarity of the generated code, instruct GPT-4 to include explanations and commentary within the code. This mimics the *chain of thought* technique, ensuring the AI provides reasoning alongside its solution.
   By embedding explanations within the code, the model generates both a solution and a tutorial-like guide for the reader.
6. **Self Reflection Invocation: Questioning the Correctness** When dealing with a complex topic or new areas, a prompt with very little information about the theory might lead GPT-4 to misunderstand which domain of knowledge the prompt refers to. One way of addressing this is to ask follow-up questions on the topic of interest to fully understand GPT-4's position. Another method is to include self-reflection comments or critique-style questions, where GPT-4 questions whether the information provided is correct or not as a form of self-reflection. This typically improves the quality of data retrieved from a file, but can also be applied to general prompts [3].
   By incorporating critiquing with self-reflection, the accuracy of text retrievals from external files can be improved, and hallucinations in responses to general prompts can be reduced.
7. **Addressing Hallucinations in Prompt Engineering** Hallucinations, a factually incorrect or fabricated content generated by language models, pose a major challenge for reliable use of AI, especially in scientific domains where accuracy is paramount [2]. Hallucinations arise from the model's training data, data sources, and inference methods, leading to errors when the model produces outputs based on memorized or low-confidence data. While end-users have limited control

over data and training-related hallucinations, errors at the inference level can be minimized by employing strategies that prioritize faithfulness to the user's prompt.
For instance, *Faithfulness Enhanced Decoding* techniques adjust the model's probability weighting to prioritize user-provided context and instruction alignment, effectively reducing hallucinatory outputs by focusing on the most relevant tokens [1]. Moreover, detailed prompts with clear and precise context reduce ambiguity, helping the model generate responses that are more likely to adhere closely to the input and reduce error. When the input prompt is sufficiently detailed, specific and clear, the likelihood of hallucinations decreases, enhancing response accuracy.
Therefore, for high-stakes scientific applications, detailed, well-structured prompts are critical. This includes iterative refinement, prompt adjustments, and, if needed, employing structured query patterns like chaining commands to guide GPT-4 in generating fully operational response.

3 Tests for Code Verification, Inbuilt and Separate

Tests for code verification are indispensable. We can distinguish testing action we already ask the Chatbot to carry out being inbuilt in the prompt or tests we figure out and design.

1. **Incorporate Testing and Verification: Request Edge Case Handling**
   Given the importance of verification, prompt GPT-4 to include testing cases, especially for edge conditions. This promotes robust code generation, essential in computational mechanics.
2. **Check Initialization of Problems**
   Several problems require an initialization of variables over a domain or an initialization of the parameters of a model. If the initialization is incorrect, the solution, despite being correct, can result in wrong results. Thus, problem initialization should be checked before moving to the verification of the rest of the solution provided by GPT-4.
3. **Verify Boundary Conditions**
   Despite clear instructions, GPT-4 may omit some boundary conditions of the problem. A simple check of the values of the variables at these boundaries can be made to ensure if they are included correctly in the problem.
4. **Ensure Data Inputs are Correct**
   For problems requiring external sources of data to be imported into the model, the correct processing of the data should be ensured. For e.g. image segmentation, if the masks are not assigned correctly, then the models will not result in the correct segmentation.
5. **Mathematical Accuracy of Numerical Implementation**
   Numerical methods applied for solving differential equations can be implemented

with minor inaccuracies which can cause major deviations from the expected outputs. For e.g., in the summation $\sum_{i \neq j}^{N} (\eta_i^n)^2$, the code generated can miss out on ensuring $i \neq j$, resulting in an erroneous summation. Such details of numerical implementation should be checked.

6. **Comparison with Analytical Solutions in Particular Settings**
Some complex problems can be reduced to a form where an analytical solution exists. Providing this information within the prompts helps GPT-4 generate the corresponding analytical equations, which can then be coded to verify the results. However, if the generated results are unsatisfactory or incorrect, providing the explicit analytical form in the form of equations will also work.
7. **Comparison with (Commercial) Software Systems** When verifying codes against commercial software, there are several steps and considerations to ensure a meaningful and robust comparison; (i) define the scope and objectives clearly (choice of a problem that can be modeled equivalently in the generated code and the commercial software; choice of comparison metrics; preference of rather simple benchmark problems), (ii) ensure model consistency (geometry, material properties, boundary conditions, loading, mesh, etc.), (iii) match solver settings (for, e.g. finite elements: element formulation, numerical integration, solver algorithm, tolerances, (iv) handle nonlinearities carefully.
Observed differences should be interpreted cautiously. Differences can arise due to hidden reasons. As an example, commercial software often uses highly optimized and proprietary algorithms that may differ in subtle ways. Moreover, slight differences in floating-point arithmetic can accumulate over iterative calculations. Moreover, we recommend to treat commercial software as a "gold standard" cautiously; their results are not inherently infallible.

4 Current Limitations–Where, Why and How GPT-4 Fails and How to Overcome It

1. **Long, Complex Coding Tasks**
Even though GPT-4 has large context windows of 128000 context length [4], for a long task they are not sufficient. There are multiple ways it can be mitigated. Some of them are,

    - Explicitly informing GPT-4 that prompt is long and to wait for complete code.
    - Break down the codes in sections (instead of splitting as per max tokens), will help GPT-4 to better understand the context, than stopping at a random point. Using self-reflections and chains of thought improves the response quality.

2. **Task Skipping**
When structuring highly complex prompts with multiple tasks that reference each other, such as "Task 1. Task 2. Task 3 refers to Task 2. Task 4 now refers to Task 3...", GPT-4 might forget some tasks, leading to their complete omission.

In cases where task skipping is evident, it is always recommended to restructure the prompts in a clearer and more straightforward fashion.

3. **Sticking to Older Library Versions**
For Python libraries that have been updated to newer versions recently, GPT-4 may provide code with the syntax and usage according to older versions. This may cause incompatibility with other libraries at times. To remedy this issue, a prompt containing the current usage of library methods from the original documentation can be added to update GPT-4 generated code.
4. ***'r'*s in *Strawberry***
One of the well-known examples of GPT-4's tokenization issue is asking, "*How many 'r's are in the word 'Strawberry'?*". Despite the obvious answer being 3, GPT-4 often provides different answers, a result of the tokenization process. GPT-4 typically reads tokens based on attention, meaning that when it processes the question, information about the word might appear before or after the prediction, and the weight of the last token might not influence the result.
Self-reflection, by prompting the same question again or questioning the correctness of the answer, might resolve the issue. Another possible approach is to add the instruction *"Compute with code."* to the prompt. This allows GPT-4 to compute the answer by generating a program that verifies and provides the correct response. However, in a recent version before book publication, this issue has been patched.

References

1. Huang, L., Yu, W., Ma, W., Zhong, W., Feng, Z., Wang, H., Chen, Q., Peng, W., Feng, X., Qin, B. & Liu, T. A Survey on Hallucination in Large Language Models: Principles, Taxonomy, Challenges, and Open Questions. *ACM Trans. Inf. Syst.*. (2024, 11)
2. Ji, Z., Lee, N., Frieske, R., Yu, T., Su, D., Xu, Y., Ishii, E., Bang, Y., Madotto, A. & Fung, P. Survey of hallucination in natural language generation. *ACM Computing Surveys*. **55**, 1–38 (2023)
3. Asai, A., Wu, Z., Wang, Y., Sil, A. & Hajishirzi, H. Self-rag: Learning to retrieve, generate, and critique through self-reflection. *ArXiv Preprint* ArXiv:2310.11511. (2023)
4. OpenAI, "OpenAI Platform - Models Documentation," 2024. [Online]. Available: https://platform.openai.com/docs/models/. [Accessed: 26- Sep- 2024].

Index

© The Editor(s) (if applicable) and The Author(s), under exclusive license to Springer Nature Switzerland AG 2025
B. Eidel (ed.), *GPT for Python-Coding in Computational Materials Science and Mechanics*, Studies in Computational Intelligence 1198,
https://doi.org/10.1007/978-3-031-85470-5

The manufacturer's authorised representative in the EU is Springer Nature Customer Service Centre GmbH, Europaplatz 3, 69115 Heidelberg, Germany. If you have any concerns regarding our products, please contact ProductSafety@springernature.com

Printed and bound by CPI Group (UK) Ltd, Croydon, CR0 4YY

30/09/2026

02225571-0006